NEXT STEPS

Bridging the gap between
Numeracy and NCEA Level 1

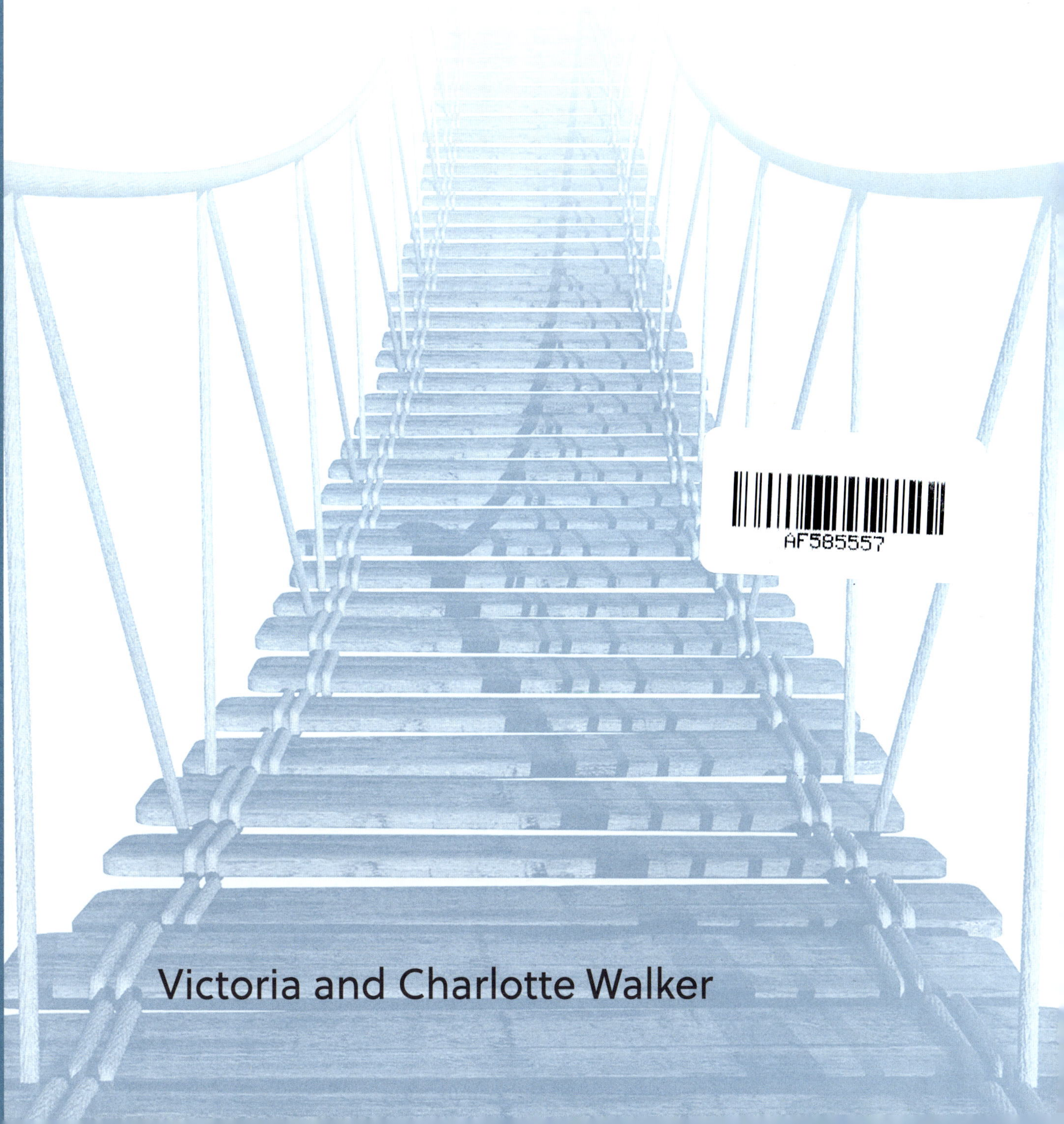

Victoria and Charlotte Walker

Walker Maths: Next Steps
1st Edition
Charlotte Walker
Victoria Walker

Cover design: Cheryl Smith, Macarn Design
Text design: Cheryl Smith, Macarn Design
Production controller: Siew Han Ong

Acknowledgements
The authors and publishers wish to acknowledge the trusted kindred spirits throughout the country for your help in preparing this title.

All images are courtesy of Shutterstock or iStock.

For product information and technology assistance,
in Australia call **1300 790 853**;
in New Zealand call **0800 449 725**

For permission to use material from this text or product, please email **aust.permissions@cengage.com**

National Library of New Zealand Cataloguing-in-Publication Data
A catalogue record for this book is available from the National Library of New Zealand.

978 0 17 048408 4

Cengage Learning Australia
Level 5, 80 Dorcas Street
Southbank VIC 3006 Australia

Printed in China by 1010 Printing International Limited.
3 4 5 6 7 28 27 26 25

CONTENTS

Factors, multiples and primes

Factors

- The **factors** of a number are all the numbers that **divide** exactly into it.
  Example: The factors of **9** are: **1**, **3**, 9.
  The factors of **24** are: **1**, 2, **3**, 4, 6, 8, 12, 24.
- The **common factor** of two numbers are those that are in both lists.
  Example: The common factors of **9** and **24** are **1** and **3**.
- The **highest common factor** (HCF) of two numbers is the **biggest factor** of both numbers.
  Example: The HCF of **9** and **24** is **3** because it is the biggest factor that is on both lists.

List all the factors of the following numbers.

1 4: ______________________

2 12: ______________________

3 15: ______________________

4 36: ______________________

5 7: ______________________

6 1: ______________________

List the factors of both numbers, and write down the HCF.

7 8: ______________________

12: ______________________

Common factors: ______________

8 5: ______________________

20: ______________________

Common factors: ______________

9 18: ______________________

60: ______________________

HCF: __________

10 11: ______________________

21: ______________________

HCF: __________

11 16: ______________________

30: ______________________

HCF: __________

12 17: ______________________

3: ______________________

HCF: __________

13 **a** Henry has 24 mint lollies and 36 fruit lollies. He wants to use all of these to make as many identical small bags of lollies as possible. How many bags can he make, and what does each contain?

____

____

b He buys an additional 30 caramel lollies. How many identical bags can he make now, and what does each contain?

____

Multiples

- The **multiples** of a number are the results of **multiplying** the number by another number.

  Example: The multiples of **3** are: 3, 6, 9, **12**, 15, 18, ...
  The multiples of **4** are: 4, 8, **12**, 16, 20, 24, ...

  These are the answers to the times tables.

- The **lowest common multiple** (LCM) of two numbers is the smallest multiple of both numbers.

  Example: The LCM of **3** and **4** is 12 because it is the smallest multiple that is on both lists.

List the first six multiples of the following numbers.

1 2: ____________

2 5: ____________

3 9: ____________

4 12: ____________

5 10: ____________

6 1: ____________

List at least the first six multiples of both numbers, and write down the LCM.

7 2: ____________
3: ____________
LCM: ______

8 5: ____________
3: ____________
LCM: ______

9 6: ____________
9: ____________
LCM: ______

10 8: ____________
12: ____________
LCM: ______

11 10: ____________
12: ____________
LCM: ______

12 7: ____________
9: ____________
LCM: ______

13 **a** Melino is threading green and white beads to make a pattern. She decides that every tenth bead will be green and every eighth bead will be a big one. Which bead will be the first big green one?

____

____

b Her green and white beads, as well as being big or small, can also be round and cubic shapes. If she does the patterns in part **a** but she makes every third bead a cubic one, which bead will be the first big, green, cubic bead?

____

____

ISBN: 9780170484084

Prime numbers

- A prime number has **exactly two factors**: 1 and itself.

Examples: 2 **is** a prime number because it has exactly **two** factors: **1** and **2**.
13 **is** a prime number because it has exactly **two** factors: **1** and **13**.
8 **is not** a prime number because it has **four** factors: **1**, **2**, **4** and **8**.

Note: 1 **is not** a prime number because it has only **one** factor: **1**.

1 Highlight the prime numbers between 1 and 50. (Hint: You should find 15.)

1	2	3	4	5	6	7	8	9	10
11	12	13	14	15	16	17	18	19	20
21	22	23	24	25	26	27	28	29	30
31	32	33	34	35	36	37	38	39	40
41	42	43	44	45	46	47	48	49	50

2 Are all prime numbers odd? Give a reason for your answer.

____

____

3 Are all odd numbers prime? Give a reason for your answer.

____

____

4 What is the next prime number after 50? ______________________________

5 Place these numbers in the correct area: 14, 31, 1, ~~7~~, 15, 19, 22, 2, 33, 37, 59.

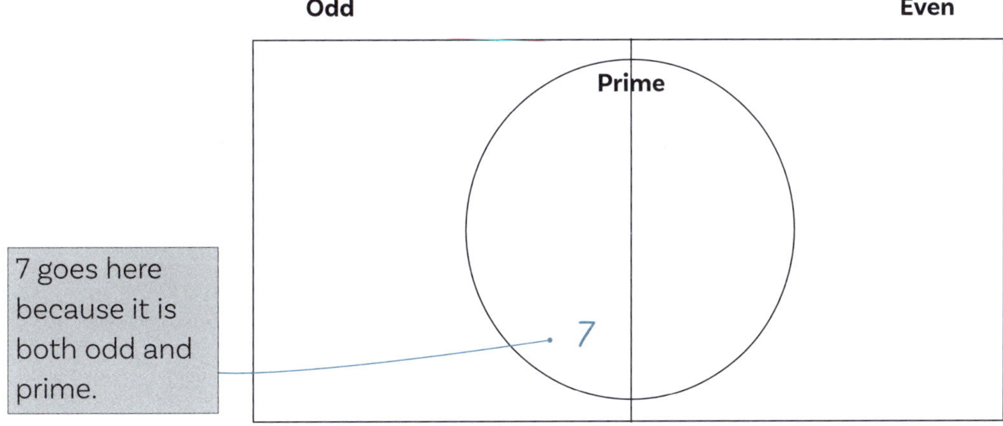

Prime factorisation

- All non-prime numbers can be written as products of prime factors.
- Prime numbers and their factors are very important in cryptography (writing and using codes), especially for use in cyber security.

Examples:

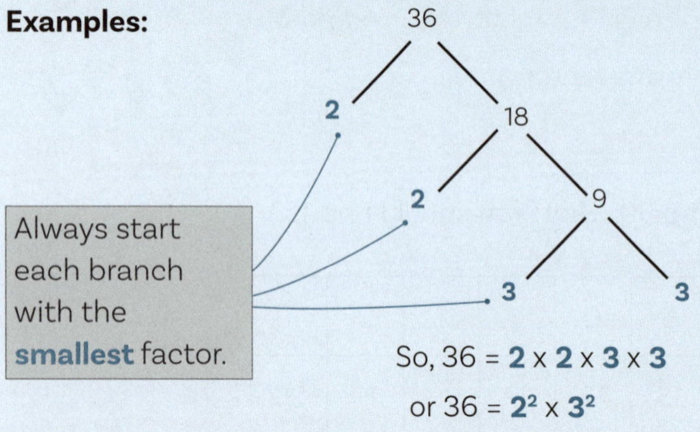

So, 36 = **2** x **2** x **3** x **3**

or 36 = $\mathbf{2^2}$ x $\mathbf{3^2}$

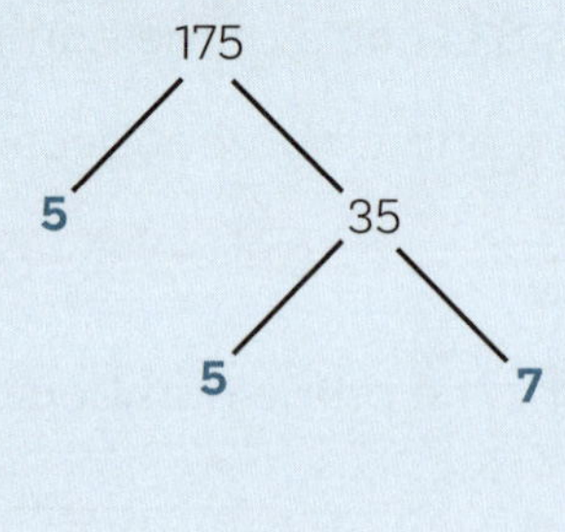

175 = **5** x **5** x **7**

or 175 = $\mathbf{5^2}$ x **7**

Complete these prime factor trees.

1

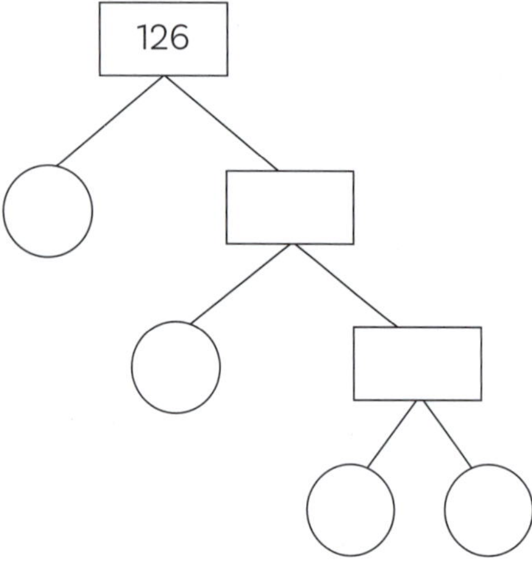

126 = ______________________

2

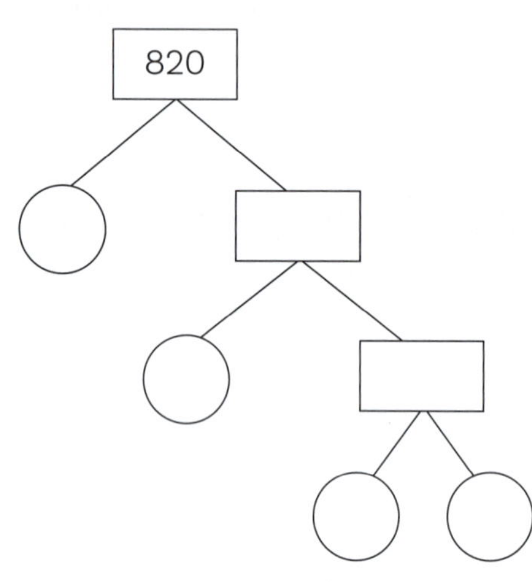

820 = ______________________

3

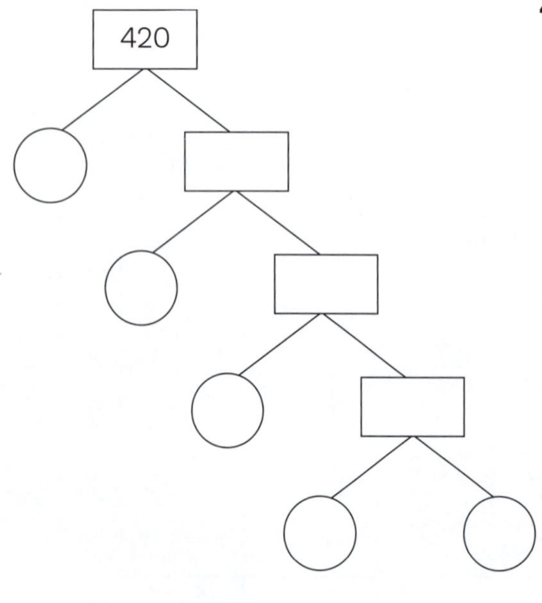

420 = ______________________

4

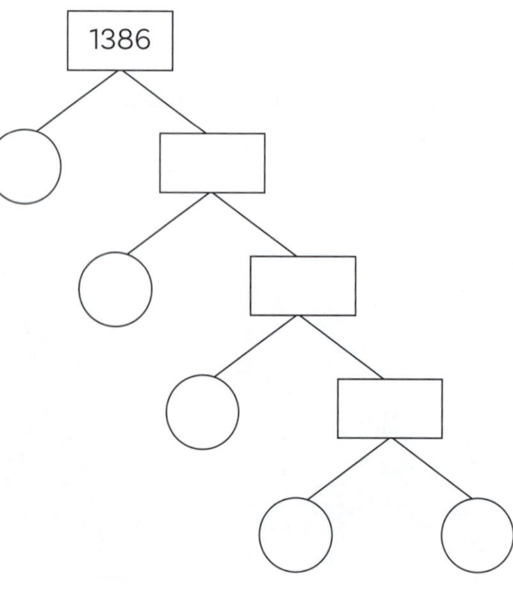

1386 = ______________________

ISBN: 9780170484084

Powers and roots

Powers

Powers are used to indicate how many times a number is multiplied by itself.

The **power** indicates how many times you need to multiply.

Make sure you use **brackets** on your calculator.

Examples: $3^2 = 3 \times 3$
$= 9$

$(-4)^3 = -4 \times -4 \times -4$
$= -64$

You can do this on your calculator with either  or $x^{\blacksquare}$

Calculate these powers.

1 6^2

2 4^4

3 $(-2)^3$

4 $(-3)^4$

5 -3^2

6 $-(-5)^3$

7 $\left(\frac{1}{2}\right)^2$

8 0.25^2

Write these numbers as powers.

9 25

10 27

11 -32

12 1000

13 1 000 000

14 0.01

ISBN: 9780170484084

Fractional powers — roots

- Finding a root is the opposite of finding a power.

You do not need to write $\sqrt[2]{x}$.

- The **square root** is written as $\sqrt{x}$ or $x^{\frac{1}{2}}$.

Example: $\sqrt{9} = \sqrt{? \times ?}$
$= \sqrt{3 \times 3}$
$= 3$

- The **cube root** is written as $\sqrt[3]{x}$ or $x^{\frac{1}{3}}$.

Example: $\sqrt[3]{-64} = \sqrt[3]{-4 \times -4 \times -4}$
$= -4$

- The **fourth root** is written as $\sqrt[4]{x}$ or $x^{\frac{1}{4}}$.

Example: $\sqrt[4]{10\,000} = \sqrt[4]{10 \times 10 \times 10 \times 10}$
$= 10$

You can do this on your calculator with 

To access this function, use the shift key. 

Calculate these fractional powers.

1 $16^{\frac{1}{2}}$ ______________________

2 $25^{\frac{1}{2}}$ ______________________

3 $125^{\frac{1}{3}}$ ______________________

4 $16^{\frac{1}{4}}$ ______________________

5 $0.25^{\frac{1}{2}}$ ______________________

6 $\left(\frac{1}{4}\right)^{\frac{1}{2}}$ ______________________

Estimating the size of a root

You can estimate the size of a square root by considering the sizes of the perfect squares on either side.

Example: Estimate $\sqrt{30}$.

$5^2 = 25$

30 lies between 25 and 36

$6^2 = 36$

$\therefore \sqrt{30}$ must lie between 5 and 6

Complete the second and third columns **without** using a calculator. Select values for the third column from this list: 8.185, 4.472, ~~6.325~~, 2.449, 8.944.

Square root	Lower and upper limits for square root	Answer
$\sqrt{40}$	*6 and 7*	6.325
$\sqrt{6}$		
$\sqrt{80}$		
$\sqrt{67}$		
$\sqrt{20}$		

ISBN: 9780170484084

Negative powers

- These are the reciprocals of positive powers.
- To find the reciprocal, you turn a fraction upside down.

Examples: $3^{-2} = \frac{1}{3^2} = \frac{1}{9}$ $\quad (-4)^{-3} = \frac{1}{(-4)^3} = -\frac{1}{64}$

$10^{-2} = \frac{1}{10^2} = \frac{1}{100}$ $\quad \left(\frac{3}{7}\right)^{-2} = \left(\frac{7}{3}\right)^2 = \frac{49}{9}$

Calculate these negative powers.

1 2^{-1}

______

2 10^{-2}

______

3 2^{-4}

______

4 10^{-6}

______

5 $\left(\frac{5}{4}\right)^{-1}$

______

6 $\left(\frac{1}{3}\right)^{-1}$

______

7 $16^{-\frac{1}{2}}$

______

8 $\left(\frac{1}{9}\right)^{-\frac{1}{2}}$

______

Mixing it up

1 $3^2 \times 2$

______

2 3×4^3

______

3 $-(-4)^2$

______

4 -4^2

______

5 $\frac{2^6}{2^2}$

______

6 $\frac{6^2}{\sqrt{9}}$

______

7 -5^{-2}

______

8 $(\sqrt{4})^2$

______

9 $3^2 + (-3)^2 - (-3)^2$

______

10 $20 + 5^3$

______

11 $\sqrt[3]{27} \times 4^2$

______

12 $(-2 - 4)^2 + (3 \times 2)^2$

______

Rounding

- Often we need to round numbers to sensible and/or meaningful values.
- Never round until **after** you have completed all calculations.
- When rounding, the decimal point never moves.

1 Rounding to decimal places

- When asked to round to 2 dp (two decimal places), this means there should be **exactly** two digits after the decimal point.

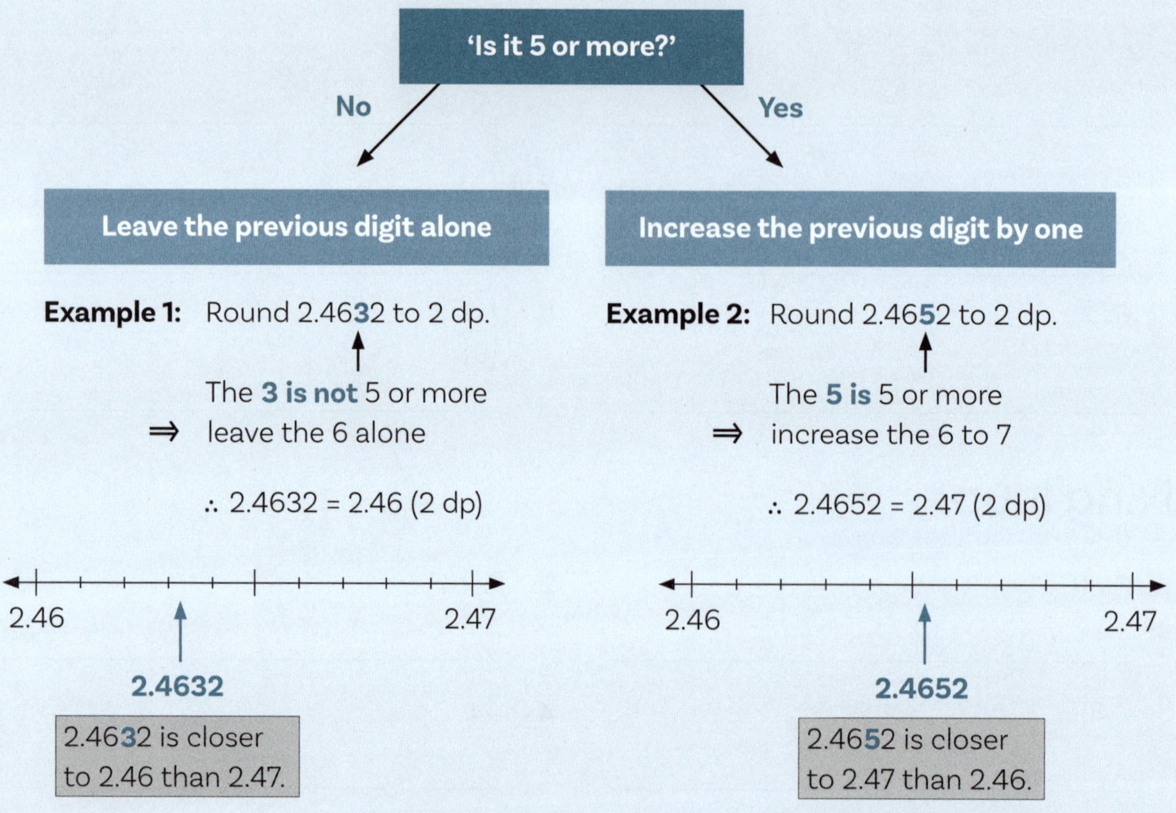

Some examples:

Unrounded number	Number of decimal places	Rounded number
0.129	2	0.13
3.824499	3	3.824
1.73051	3	1.731
0.999	1	1.0
4.0962	2	4.10
699.501	0	700

Your answer **must** have the required number of decimal places, so don't omit the 0s.

ISBN: 9780170484084

Round these to the required number of decimal places.

1 6.82331 (2 dp)

2 2.467981 (2 dp)

3 64.1302 (1 dp)

4 142.09 (1 dp)

5 9.0912 (3 dp)

6 91.4458 (3 dp)

7 399.899 (2 dp)

8 0.0314 (2 dp)

9 0.004 (2 dp)

10 1.0203 (2 dp)

11 2.99 (1 dp)

12 2.36^2 (1 dp)

13 $\sqrt{89}$ (2 dp)

14 86.3 x 23.5 (0 dp)

15 $\sqrt[3]{568}$ (1 dp)

16 $\frac{4.378 \times 5.699}{\sqrt{76.592}}$ (3 dp)

17 A patient has their temperature taken. It is 38.16°C.

a Round the temperature to the nearest degree (i.e. 0 dp).

b Round it to the nearest tenth of a degree (i.e. 1 dp).

18 A rugby field is 103.41 m long and 49.26 m wide.

a Calculate its area to the nearest square metre.

b The groundsman only needs to know the area to the nearest 10 m^2 (for fertilising). Round the area to the nearest 10 m^2.

2 Rounding to significant figures

Rules for deciding whether a digit is significant

1 All non-zero digits are significant.

2 Zero **is** significant if it is

- between non-zero digits — **Examples:** 3105 has 4 significant figures; 1.0032 has 5 significant figures
- at the end of a decimal. — **Examples:** 2.70 has 3 significant figures; 16.00 has 4 significant figures

3 Zero **is not** significant if it is

- at the start of a number smaller than one — **Examples:** 0.005 has 1 significant figure; 0.592 has 3 significant figures
- at the end of a number which has no decimal point. — **Examples:** 2840 has 3 significant figures; 100 has 1 significant figure

Write down the number of significant figures in the following.

Number	Number of significant figures
51.2	
13 706	
14.028	
94 778.5	
10	
3.20046	
4 461 030	
12 000	
1.20	
1205	
0.2	

Number	Number of significant figures
0.004	
43.50	
9.670420	
53.060	
2.0670	
79 000 001	
10 000	
60.0850	
0.060140	
1.030	
0.090	

ISBN: 9780170484084

How to round to significant figures

- When asked to round to 4 sf (four significant figures), this means there should be exactly four significant figures in the answer.
- The process is almost the same as rounding to decimal places.
- **Always** check that the rounded number is a similar size to the original number.

Locate the digit to the **right** of the last required significant figure.

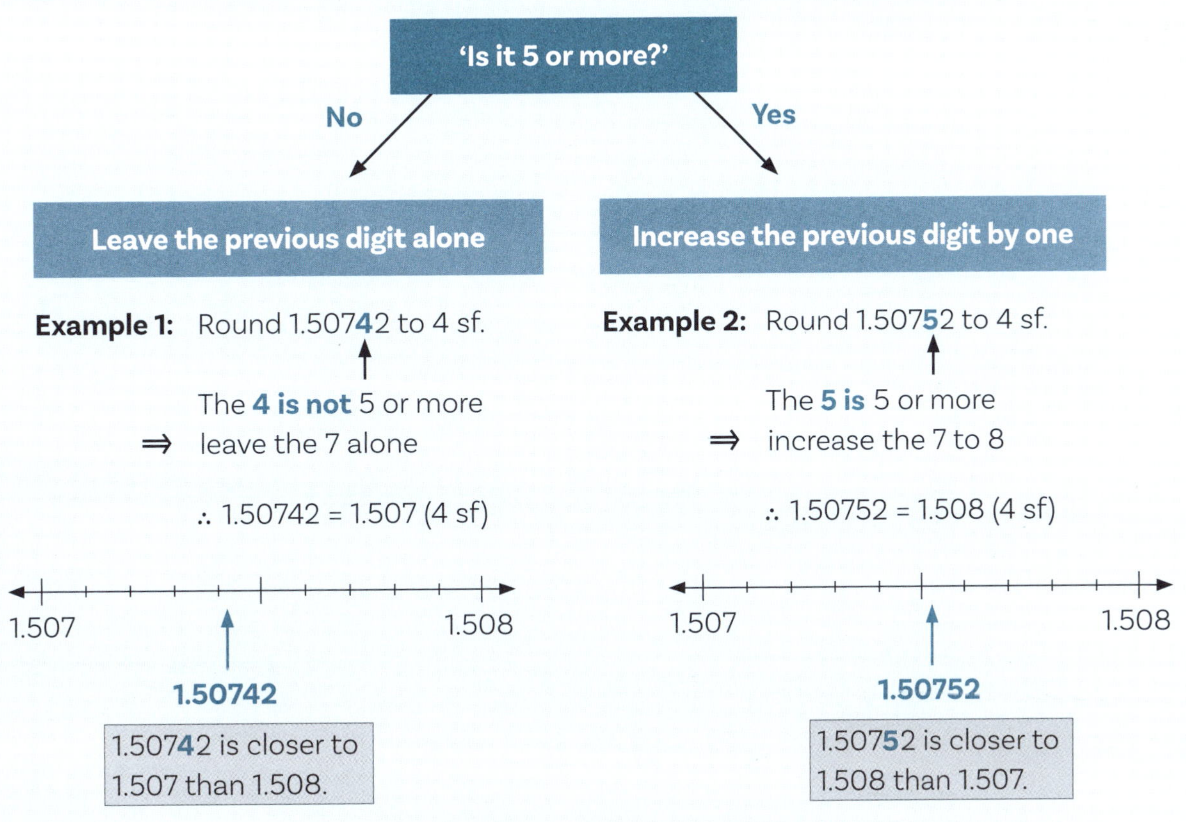

Some examples:

Unrounded number	Number of significant figures	Rounded number
123 456	2	120 000
4.26794	4	4.268
17.500	2	18
0.12079	4	0.1208
0.020789	4	0.02079
35.7051	4	35.71
29.5	2	30

Sometimes when you round to 2 sf, you get a number that looks as though it is rounded to 1 sf.

Unrounded number	Number of significant figures	Rounded number
67 291	4	67 290
0.999	2	1.0
4.0962	3	4.10
0.0038962	3	0.00390
699.501	1	700
109.9	3	110
1.795	3	1.80

Your answer **must** have the required number of significant figures, so don't omit the 0s.

Round these to the required number of significant figures.

1 52.8 (2 sf)

2 34 651 (2 sf)

3 0.0678 (2 sf)

4 10 462 (2 sf)

5 109.8 (3 sf)

6 0.001055 (3 sf)

7 741 980 (3 sf)

8 0.0069421 (3 sf)

9 1 671 512 (4 sf)

10 0.069995 (4 sf)

11 67.8 (1 sf)

12 94 267 (1 sf)

13 0.47911 (1 sf)

14 0.0555 (1 sf)

15 0.10683 (1 sf)

16 0.00789 (1 sf)

17 $\sqrt{0.887}$ (3 sf)

18 0.125 x 0.863 (3 sf)

19 1.522 ÷ 0.7 (3 sf)

20 0.563 + 0.679 + 0.441 + 0.105 (3 sf)

21 If 11 and 12 have both been rounded to 2 sf, calculate the lowest and highest possible values for the sum of the unrounded numbers.

 ISBN: 9780170484084

Rates

- A rate is a relationship between two quantities.
- Rates are (often) written using the word '**per**'.
- **Per** means 'for each' and is another way of saying **divide**.

  Examples: the amount paid per hour for babysitting
  kilometres per hour, the speed at which a car travels.
- When we use two rates in a calculation, we assume that the relationship between them remains constant. They stay in proportion.

Example 1: 3 kg of carrots cost $4.95. Calculate the cost of 5 kg of carrots.

Method 1

Step 1: Write the information you are given in a brief statement:

3 kg cost **$4.95**

Because you are trying to find a price, put the **price** on the right.

Step 2: Underneath, write a parallel statement about what you want to know:

3 kg cost **$4.95**

copy ↓

5 kg cost **$4.95** x $\frac{5}{3}$ = $8.25

Multiply by either $\frac{3}{5}$ or $\frac{5}{3}$. Use $\frac{5}{3}$ here because the $4.95 needs to be made **bigger**.

or

Method 2

Step 1: Work out how much 1 kg costs.

1 kg of carrots costs **$4.95** ÷ 3 = $1.65

Step 2: Multiply the cost of 1 kg by the number of kilograms you need.

5 kg of carrots cost **$1.65** x 5 = $8.25

Example 2: Sinta sometimes walks to school, which takes 27 minutes at a speed of 4 kph. Otherwise she skateboards to school at 10 kph. How long should skateboarding to school take her?

Step 1: Write the information you are given in a brief statement:

At 4 kph she takes **27 min**

Because you are trying to find a time, put the **time** on the right.

Step 2: Underneath, write a parallel statement:

At 4 kph she takes **27 min**

copy ↓

At 10 kph she takes **27** x $\frac{4}{10}$ = 10.8 min

Multiply by either $\frac{4}{10}$ or $\frac{10}{4}$. Use $\frac{4}{10}$ here because the 27 min needs to be made **smaller**.

Answer the following questions.

1 9 kg of apples cost $24.75. Calculate the cost of 5 kg of apples.

2 3 kg of mince cost $32.85. Calculate the cost of 7 kg of mince.

3 When Lizzie drove from home to town at an average of 48 kph, it took her 21 minutes. Her mum drove from home to town at an average of 42 kph. How long did it take her?

4 When Lizzie drove from home to the beach at an average of 72 kph, it took her 28 minutes. Her mum did the same trip in 24 minutes. What was her mum's average speed?

5 An ant has mass 0.0003 kg and can lift 0.015 kg. Adam has mass 72 kg. If he could lift the equivalent of an ant, what mass could he lift?

6 On average, people speak for just 10 minutes per day. If a person is awake for 16 hours in a day, on average, how long do they spend speaking during each hour.

7 A snail can crawl 49 m in one hour. How far can it crawl in one minute?

8 In the USA, 2.5 cans of spam are consumed per second. How many cans of spam are consumed in the USA in one hour?

 ISBN: 9780170484084

Exchange rates

- The value of the New Zealand dollar changes in relation to other currencies and with time.
- Changes in its value are important when calculating the cost of overseas goods or travel.
- You can use the same method as in proportion calculations.
- Currency units: the New Zealand dollar can be written as $NZ or NZD.

Exchange rates used in the examples and exercises: $NZ1 = €0.65 (euro)
$NZ1 = £0.49 (pounds sterling)
$NZ1 = $US0.74 (US dollar)

Example 1: Michelle takes $3000 spending money with her to the United States.
How many $US is that worth?

This could also be written as NZ$1.

$$\text{\$NZ1} = \text{\$US0.74}$$
$$\therefore \text{\$NZ3000} = \text{\$US0.74} \times \frac{3000}{1}$$
$$= \text{\$US2220}$$

This could also be written as US$0.74.

Example 2: Ali returns to New Zealand with €8500 earned while working in the Netherlands.
How many $NZ will he get when he banks it?

$$€0.65 = \text{\$NZ1}$$
$$\therefore €8500 = \text{\$NZ1} \times \frac{8500}{0.65}$$
$$= \text{\$NZ13 076.92}$$

Solve the following exchange rate problems.

1 Convert $NZ2500 to €.

2 Convert $NZ2500 to $US.

3 Convert $NZ2500 to £.

4 Convert €150 to $NZ.

5 Convert $US600 to $NZ.

6 Convert £15 to $NZ.

ISBN: 9780170484084

Ratios

Simplifying ratios

- Ratios show how an amount is split into several shares, usually of different sizes.
- Before you convert quantities into ratios, both quantities should be in the **same units**.
- Ratios are usually given in whole numbers, not decimals or fractions.
- A **colon** (**:**) is used to separate the parts, e.g 2**:**1 means 'two **to** one'.
- Like fractions, ratios should be written in their simplest form.

Remember: Highest common factor (HCF) is the biggest number that will divide into several numbers. For example, the HCF of 15 and 12 is 3.

Example 1: Write the ratio 60:36:24 in its simplest form.

$$60 : 36 : 24 = 5 : 3 : 2$$

Divide each of these numbers by their HCF (**12**).

These numbers have no common factors.

Example 2: Write the ratio $1\frac{1}{2} : \frac{2}{3} : 2$ in its simplest form.

$$1\frac{1}{2} : \frac{2}{3} : 2$$

$$1\frac{1}{2} \times \mathbf{6} : \frac{2}{3} \times \mathbf{6} : 2 \times \mathbf{6}$$

$$9 : 4 : 12$$

Multiply each of these numbers by a number big enough to get rid of the fractions (**6**).

Example 3: Write the ratio 0.5 : 0.15 : 0.2 in its simplest form.

$$0.5 : 0.15 : 0.2$$

$$0.5 \times \mathbf{100} : 0.15 \times \mathbf{100} : 0.2 \times \mathbf{100}$$

$$50 : 15 : 20$$

$$10 : 3 : 4$$

Multiply each of these numbers by a number big enough to get rid of the decimal point (**100**).

Then divide each number by its HCF (**5**).

Write the following ratios in their simplest forms.

1 6 : 4

____

2 100 : 70 : 30

____

3 16 : 20

____

4 $\frac{1}{2} : \frac{3}{4}$

____

5 $3\frac{1}{2} : 2\frac{1}{2}$

____

6 $\frac{1}{2} : \frac{1}{3} : \frac{1}{4}$

____

ISBN: 9780170484084

7 0.1 : 0.5

____

8 1.2 : 3.6 : 2.4

____

9 0.04 : 0.06 : 0.12

____

10 12 m : 0.5 m

____

11 1 hour : 10 min

____

12 2 km : 3.5 km : $\frac{1}{2}$ km

____

Ratio calculations where the total amount is given

Example 1: \$45 needs to be split between two people in the ratio 2:3.

Step 1: Calculate the **total number of parts** by adding the numbers in each share: 2 + 3 = **5**

Step 2: Calculate the **value of each part** by dividing the total amount by the number of parts: \$45 ÷ 5 = **\$9**

Step 3: Calculate the **value of each share** by multiplying the number of parts by the value of each part: 2 x \$9 : 3 x \$9

∴ \$18 : \$27

One person gets \$18 and the other gets \$27. (Check: \$18 + \$27 = \$45)

Example 2: Annie, Amira and Amelia paint their grandma's fence. She will pay them a total of \$180. Annie works for five hours, Amira for four hours, but Amelia gets bored and leaves after three hours. How much should each girl get paid?

Step 1: Calculate the **total number of parts** by adding the numbers in each share: 5 + 4 + 3 = **12**

Step 2: Calculate the **value of each part** by dividing the total amount by the number of parts: \$180 ÷ 12 = **\$15**

Step 3: Calculate the **value of each share** by multiplying the number of parts by the value of each part: 5 x \$15 : 4 x \$15 : 3 x \$15

∴ \$75 : \$60 : \$45

Annie gets \$75, Amira gets \$60 and Amelia gets \$45.
(Check: \$75 + \$60 + \$45 = \$180)

Solve the following ratio problems.

1 Sam and Sarah pool their money to buy a \$3 bag of lollies. Sam gave \$2 and Sarah gave \$1. If there are 51 lollies in the bag, how many lollies should each get?

____

____

2 A recipe requires two parts of liquid for every three parts of dry ingredients. If Hazel wants to make a total of $2\frac{1}{2}$ cups, how much of each type of ingredient will she need?

____

____

ISBN: 9780170484084

3 A cordial bottle states that one part of cordial is needed for every seven parts of water. What volumes of cordial and water are needed to make up 24 L of drink for the school social?

---

4 The human body contains 1.3 bacterial cells for every one human cell. If the total number of human and bacterial cells in a body is 69 trillion (10^{12}), calculate the number of bacterial cells and human cells.

---

5 The ratio of people with the normal number of ribs to those with one extra rib is 20:1. In a town of 13 272 people, how many would be expected to have an extra rib?

---

6 Butter and cheese are produced in New Zealand in a mass ratio of 5:3. If a total of 864 000 tonnes of butter and cheese was produced last year, how much of each was produced?

---

7 Uncle Scrooge revises his will. He leaves his $38 000 in the ratio of four parts to the cat, one part to each of the four older nieces, and two parts to his favourite niece, the youngest. How much does each get?

---

8 The town plan requires that the ratio of floor area to total land area in a housing development is less than 0.8:1. If a section is 535 m^2, calculate the maximum floor area that a house can have.

---

9 The town plan also requires that new subdivisions have a quarter of the land area set aside as open space. The rest is to be used in the ratio of four parts high-density housing to three parts medium-density housing to one part low-density housing. If the area of a new subdivision is 56 ha, calculate the areas used for open space and each type of housing.

ISBN: 9780170484084

Ratio calculations where one part is given

Example 1: The ratio of people to cars in New Zealand is about 8:5. If there are 5 123 000 people, about how many cars are there?

Step 1: Calculate the **value of each part** by dividing the value given (5 123 000) by the number of parts (8) it represents: 5 123 000 ÷ 8 = **640 375**

Step 2: Calculate the **value of the share** by multiplying the number of parts (5) by the value of each part: 5 x 640 375 = 3 201 875

∴ there are about 3 201 875 cars in New Zealand.

Example 2: An orchard is being planted with apple and pear trees in a ratio of 6:5. If 174 apple trees have been planted, how many trees will be planted in total?

Step 1: Calculate the **value of each part** by dividing the value given (174) by the number of parts (6) it represents: 174 ÷ 6 = **29**

Step 2: Calculate the **value of the total amount** by multiplying the number of parts (11) by the value of each part: 11 x 29 = 319

∴ there will be 319 trees in the orchard.

Solve the following ratio problems.

1 There are about five sheep for every person in New Zealand. If the population is now 5 123 000, about how many sheep are there in New Zealand?

____

____

2 The ratio of growth in length of fingernails to toenails is 4:1. If my fingernails grew 5.6 mm in the last month, how much did my toenails grow?

____

____

3 The ratio of dried fruit to nuts in a recipe is 5:2. If Susie has 140 g of nuts, how much dried fruit will she need?

____

____

4 The ratio of boys to girls in a class is 4:3. If there are 16 boys in the class, how many students are there in total?

5 Caterpillars have 4000 muscles. The ratio of muscles in caterpillars to muscles in humans is 20:3. How many muscles do humans have?

6 In the athletic sports, the ratio of entries in running, jumping and throwing events was 8:3:2. If there were 296 entries into the running events, how many entries were there in total?

7 In the movie *The Wizard of Oz*, Judy Garland was paid $500 per week. The ratio of her salary to that of Toto, the dog, was 25:3. How much was the dog paid?

8 Moths and butterflies are called lepidoptera. The ratio of moth species to butterfly species is 35:6. If there are 24 000 species of butterflies, how many species of lepidoptera are there?

9 When tossing a coin, the ratio of tails to heads is actually 500:495 (the picture of the head weighs more, so it ends up on the bottom more often). Ethan tossed a coin until he got 200 tails. Estimate the total number of times that he tossed his coin.

10 The scale on a map states 1:50 000. If a section of track on the map is 4.9 cm long, how long is the real section of track (in m)?

11 A model car is made in the ratio 1:64. If the real car is 4.0 m long, how long is the model car (in cm)?

ISBN: 9780170484084

Scale diagrams

- The scale tells us the ratio of a length in the picture or map compared with the corresponding length of the real object or the distance.
- Notice that the lengths showing a scale **must be in the same units**.

Example 1: A rectangular paddock is shown on a map:

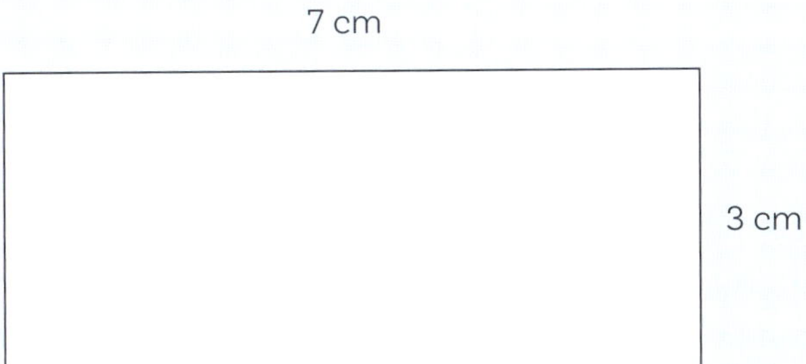

If the actual length of the short side of the paddock is 96 m, calculate the length (L) of the long side.

$$\frac{\text{short side on the map}}{\text{short side on the ground}} = \frac{\text{long side on the map}}{\text{long side on the ground}}$$

$$\frac{3}{96} = \frac{7}{L}$$

$$3L = 7 \times 96$$

$$L = \frac{7 \times 96}{3}$$

$$L = 224 \text{ m}$$

Example 2: Rua has a poster of this car. On the poster, the car is 865 mm long and 270 mm high. He wants to make a scale drawing of the car. His drawing needs to be exactly 30 mm high. How long will it be?

$$\frac{\text{height on the poster}}{\text{height on the drawing}} = \frac{\text{length on the poster}}{\text{length on the drawing}}$$

$$\frac{270}{30} = \frac{865}{L}$$

$$270L = 30 \times 865$$

$$L = \frac{30 \times 865}{270}$$

$$L = 96.1 \text{ mm}$$

ISBN: 9780170484084

Calculate the unknown lengths.

1 The diagram represents a rectangular paddock as it is shown on a map. The longer side of the paddock is 90 m. Use your measurements from the diagram to help you calculate the length of the shorter side (S) of the paddock.

90 m

S

2 Melanie is helping to make the set for the school production. The director wants a 2.1 m high version of the soldier. Measure his height on the picture, and use this to help you calculate his maximum width.

3 Sora is making a model plane. She needs to calculate the width of a wing. The length of the model wing will be 225 mm. Measure the length of wing on the picture, and use this to help you calculate the width on the model wing.

4 The wingspan of a monarch butterfly is 10 cm. Measure the wingspan on the scale diagram, and use it to help you calculate the length of the body (including the head) of a monarch butterfly.

ISBN: 9780170484084

Percentages

Using percentages

Examples:

1 Finding a percentage

Find 8% of 96.

Remember, '**of**' means **x**.

You could: convert the percentage to a decimal: 8% **of** 96 = 0.08 **x** 96

= 7.68

Or: use the % function on your calculator: 8% **x** 96 = 7.68

Not all calculators are the same, so you will need to experiment until you find how yours works.

2 Increasing by a percentage

Increase 125 by 14%.

This means that we need 100% plus 14%, or 114%.

You could use your % button on your calculator.

Increased amount = 125 + (14% of 125)
= 125 + 17.5
= 142.5

or

Increased amount = 125 x 114%
= 125 x 1.14
= 142.5

3 Decreasing by a percentage

Decrease 180 by 26%.

This means that we need 100% minus 26%, or 74%.

Decreased amount = 180 – (26% of 180)
= 180 – 46.8
= 133.2

or

Decreased amount = 180 x 74%
= 180 x 0.74
= 133.2

4 Calculating a percentage

Forty-seven out of sixty-four Year 11 girls play sport. What percentage play sport?

$$\text{Percentage} = \frac{47}{64} \times 100$$

$$= 73.4\% \text{ (1 dp)}$$

Calculate these.

1 29% of 265 km = ____________

2 45% of 68 g = ____________

3 2% of 78 kg = ____________

4 72% of $1260 = ____________

5 10.5% of 48 mL = ____________

6 3.5% of $5000 = ____________

ISBN: 9780170484084

7 Increase 512 m by 84% = ______________

8 Decrease $105 by 35% = ______________

9 Increase 48 g by 46% = ______________

10 Increase 2 min by 60% = ______________

11 Decrease 86 cm by 16% = ______________

12 Decrease 55 cm by 92% = ______________

Answer the following questions.

13 Nine out of the 20 pieces of fruit in a bowl were bananas. What percentage were bananas?

____

14 In a school, 25% of 312 students bring lunch from home. How many students bring lunch from home?

____

15 A website advertises '35% discount on all items'. Calculate the discounted price of a phone which is normally $599.

____

16 A muesli recipe requires 6 cups of dry ingredients, 1 cup of which is bran. What percentage of the dry ingredients is bran?

____

17 Eru was doing a survey of vehicles entering the school. Of the 38 vehicles, 22 were private cars, 9 were motor bikes, 4 were taxis and 3 were delivery vans. Calculate the percentage of each type of vehicle.

____

18 Uncle Scrooge revised his will yet again. The new one states that 40% goes to the cat. The remainder is to be shared equally between his five nieces. If his estate is worth $38 000, calculate how much the cat inherits, and how much each niece inherits.

____

19 Researchers in America have found that in each 1 L of bottled water, there are about 216 000 nanoplastic particles. These make up 90% of plastic fragments in each bottle. How many plastic fragments did they find in total in each bottle?

____

ISBN: 9780170484084

Calculating percentage changes

Example 1: Huia's pay is increased from $14.50 to $14.79 per hour. Calculate the percentage increase.

Step 1: **Subtract** the amounts: $14.79 - $14.50 = $0.29

Step 2: **Divide** the difference by the **original** quantity and multiply by 100: $\frac{0.29}{14.50} \times 100 = 2\%$

Example 2: Matiu bought a shirt in a sale. The original price was $40, but it was reduced to $31.20. By what percentage was it reduced?

Step 1: **Subtract** the amounts: $40.00 - $31.20 = $8.80

Step 2: **Divide** the difference by the **original** quantity and multiply by 100: $\frac{8.80}{40.00} \times 100 = 22\%$

Calculate the percentage changes below.

1 From 18 to 18.9

2 From 1200 to 1080

3 From 0.480 to 0.552

4 From 720 to 540

5 From 15 kg to 15.3 kg

6 From 7600 to 7524

7 From 12 to 27

8 From 135 to 18.9

9 Louisa bought a car for $3600, but had to sell it six months later. She got $2736. Calculate her percentage loss on the car.

10 The population of an island increased from 15 864 to 17 843. Calculate the percentage increase in the population.

GST

- **GST** stands for **G**oods and **S**ervices **T**ax.
- It is added to everything you purchase and it goes to the Government to fund the running of the country.
- The standard current GST rate on all products in New Zealand is **15%**.
- Don't forget to add units and round when necessary.

Adding GST

Example: The GST-exclusive price is $95. Calculate the GST-inclusive price.

GST-inclusive amount = $95 x **1.15**
= $109.25

Convert the percentage to a **decimal and add 1**.

Add GST to these prices.

1 $86

2 $15.50

3 $89.95

4 $102.32

5 $4056.99

6 $56.75

7 $0.23

8 $65.20

9 $530.00

10 The pre-GST price of a car is $3500.

a Calculate its GST-inclusive price.

b How much is the GST on the car?

11 The pre-GST price of an ice cream is $3.47.

a Calculate its GST-inclusive price.

b How much is the GST on the ice cream?

 ISBN: 9780170484084

Calculating the pre-GST price and GST

You may be given the GST-inclusive price and asked to calculate a pre-GST price or how much GST is being paid on an item.

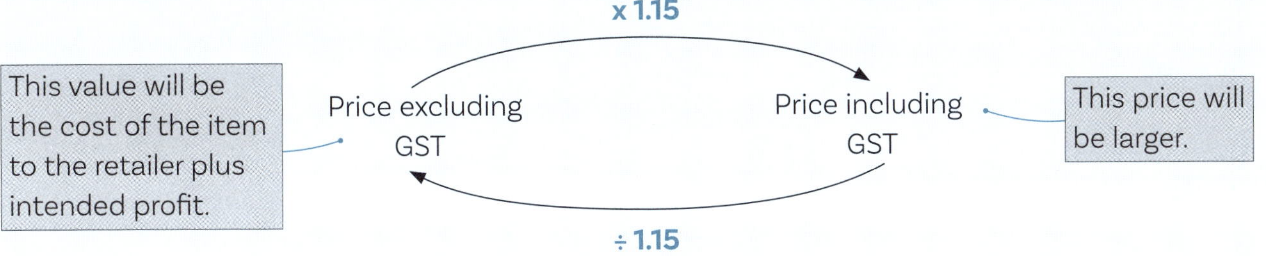

Example: A painting costs $840 including GST. How much GST was paid?

GST-exclusive price = $840 ÷ 1.15
= $730.43 (2 dp)

This is the price without GST.

∴ GST = $840 - $730.43
= $109.57

Calculate GST-exclusive prices for the following.

1 $98

2 $27.45

3 $109.99

4 $256.78

5 $0.99

6 $980.65

7 $1.59

8 $86.70

9 $890.00

10 Derek bought a car for $4999. How much GST was included in this price?

11 How much GST did Amanda pay on her ice cream that cost her $3.45?

12 A car yard had a sale. It advertised that it would pay the GST on all cars sold. The salesman subtracted 15% from the full GST-inclusive price of $5999 in order to calculate the sale price. Explain why his boss was angry, and calculate what the sale price should have been.

ISBN: 9780170484084

Interest

- When you deposit money in the bank, the bank pays you ‘interest’ for the use of that money.
- When you borrow money, you have to pay ‘interest’ because you are using the bank’s money.
- Interest may be simple interest (SI) or compound interest (CI).

Simple interest (SI)

- The same amount of interest is paid to you at regular intervals.
- This interest is paid separately into another account: it is **not** added to your deposit.

Example: Rawiri’s grandmother has lent him $4500 to buy a car. Rawiri will pay her 5% of $4500 in interest each year.

a How much does he pay each year?

Interest = 5% of $4500
= $225

b If he pays all the money back after three years, how much interest will he have paid to his grandmother in total?

Interest = 3 x $225
= $675.00

Answer the following questions. All interest rates are annual.

1 Calculate the simple interest earned when $1500 is invested at 7% for five years.

____

2 Calculate the simple interest earned when $55 000 is invested at 4.5% for 10 years.

____

3 Calculate the simple interest earned when $8750 is invested at 8% for two years.

____

4 Calculate the rate of interest needed in order to earn $66 per year from a $1200 loan.

____

5 Calculate the rate of interest needed in order to earn $1562.50 per year from a $25 000 loan.

____

6 How long will $4000 need to be invested at 6.5% in order to earn $1040 in interest?

____

ISBN: 9780170484084

Compound interest (CI)

- With compound interest, the interest is added to the amount deposited.
- The following year, interest is calculated on the total (deposit and previous year's interest).

Example: Rawiri's grandmother lends him \$4500 to buy a car, and she charges him 5% per year compound interest. This means he doesn't pay her anything annually, but pays back the \$4500 plus compound interest after three years.

a How much would he pay back at the end of the three years?

At the end of the 1st year he owes \$4500 + \$4500 x 5% = \$4725

At the end of the 2nd year he owes \$4725 + \$4725 x 5% = \$4961.25

At the end of the 3rd year he owes \$4961.25 + \$4961.25 x 5% = \$5209.31

You can do this on your calculator by entering
\$4500 x 1.05 x 1.05 x 1.05 = or \$4500 x $(1.05)^3$ =

b Compare this with the amount he would pay his grandmother when the money earned simple interest (page 32).

Simple interest: He would pay his grandmother \$4500 + \$675 interest = \$5175.

Compound interest: He would pay his grandmother \$5209.31.

∴ he would pay her \$34.31 more if she charged him compound interest.

Answer the following questions.

1 \$7500 is invested at 6% compound interest. Calculate its value at the end of two years.

2 \$25 000 is invested at 6.5% compound interest. Calculate its value at the end of three years.

3 What is the value of \$65 000 invested at 4.5% compound interest at the end of 10 years?

4 **a** When Jane was born, her grandfather invested \$2000 at an interest rate of 4.5%. Jane cannot use this until she is 18. How much will it be worth by then?

b How much more would it have been worth if it had been invested at 5%?

Standard form

- Standard form is a way of writing either very large or very small numbers without writing all the place holders (0s).
- Standard form is also known as **scientific notation**.
- A number written in standard form must look like this:

4.56×10^n

There must be exactly **one** non-zero number before the decimal point.

A **times** sign.

10 to the power (***n***) of whatever power is needed to give it the same value as the number in ordinary form.

Examples: 5 430 000 = 5.43×10^6 (Ordinary form = Standard form)

0.0000543 = 5.43×10^{-5}

Complete the following table.

Power of ten	Fraction	Whole number or decimal
10^3		
	$\frac{1}{10}$	
		1 000 000
10^{-2}		
		0.0001
	$\frac{1}{100\,000\,000}$	
10^0		

Standard form to ordinary numbers

Numbers bigger than 1

Example: Convert 7.12×10^5 to an ordinary number.

Method 1: 7.12×10^5 = 7 1 2 0 0 0 . = 712 000

5 ⇒ shift the decimal point **5** places to the **right**.

Method 2: Use your calculator: enter 7.12  **5** = ⇒ 712 000

 ISBN: 9780170484084

Numbers smaller than 1

Example: Convert 7.12×10^{-4} to an ordinary number.

Method 1: 7.12×10^{-4} = 0 . 0 0 0 7 . 1 2 = 0.000712

-4 ⇒ shift the decimal point **4** places to the **left**.

Method 2: Use your calculator: enter 7.12 [$\times 10^x$] **-4** = ⇒ 0.000712

Means '**x** 10 to the power'. (Do **not** enter the **x** sign.)

Note: For negative powers, the exponent number matches the total number of zeros:

7.12×10^{-4} = **0.000**712

Convert these to ordinary numbers.

1 9.4×10^{2} = __________

2 1.42×10^{5} = __________

3 2.69×10^{3} = __________

4 6.111×10^{6} = __________

5 8.0×10^{0} = __________

6 4.5×10^{-2} = __________

7 9.467×10^{-1} = __________

8 3.7×10^{-5} = __________

9 6.2×10^{-3} = __________

10 1.83×10^{-8} = __________

Ordinary numbers to standard form

Numbers bigger than 1

Example: Convert 634 000 to standard form.

Step 1: Shift the decimal point to the **left** until there is exactly **one** non-zero digit on its left: **6** . 3 4 0 0 0 ⇒ **6**.34

(This will give you a number between 1 and 10)

Step 2: Multiply by 10 to the power of however many places the decimal point was moved (**5**): 6.34×10^{5}

ISBN: 9780170484084

Numbers smaller than 1

Example: Convert 0.00068 to standard form.

Step 1: Shift the decimal point to the **right** until there is exactly **one** non-zero digit on its left: 0 0 0 0 **6** . **8** ⇒ 6.8

(This will give you a number between 1 and 10)

Step 2: Multiply by 10 to the **negative** power of however many places the decimal point was moved (**-4**): 6.8×10^{-4}

Check your answers by reversing the process.

Remember: To be in standard form, a number **must** be in the format **a.b x 10**^{something}

For instance: **7.4** in standard form is **7.4×10^0**

Unless b = 0
e.g. 5 in standard form is 5×10^0

Convert these to standard form.

1 543 = ______

2 1200 = ______

3 74.2 = ______

4 1.689 = ______

5 7 673 000 = ______

6 80 050 = ______

7 366 600 000 = ______

8 1 = ______

9 0.51 = ______

10 0.00067 = ______

11 0.014 = ______

12 0.000007 = ______

13 0.0001 = ______

14 0.00832 = ______

15 0.001101 = ______

16 0.0000000004 = ______

ISBN: 9780170484084

Simplifying expressions

Multiplying and dividing

The order within simplified expressions should be:

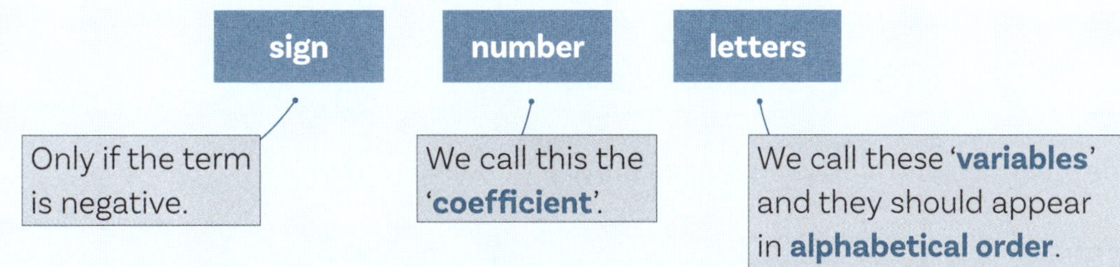

Examples:

Unsimplified expressions	Simplified expressions
$a \times 2$	$2a$
$a \times a \times a$	$\mathbf{a^3}$
$3a \times 2a$	$\mathbf{6a^2}$
$-5 \times -a \times b$	$5ab$
$a \times d \times 2c \times -4b$	$-8abcd$
$a \div 4$	$\frac{a}{4}$ or $\frac{1}{4}a$

Index form or **power form**.

Simplify the following expressions.

1 $y \times y \times y =$ ______________

2 $d \times f \times a =$ ______________

3 $7 \times d =$ ______________

4 $2d \times 3f =$ ______________

5 $-e \times 3f \times g =$ ______________

6 $d \times f \times f \times 7 =$ ______________

7 $5a \times -2a =$ ______________

8 $-4b \times 2a \times 3c =$ ______________

9 $5dc \times -d =$ ______________

10 $-1 \times 3f \times -6 =$ ______________

11 $g \div 2 =$ ______________

12 $14d \div e =$ ______________

13 $3 \div e =$ ______________

14 $20dg \div f =$ ______________

14 $(3d \times 4f) \div ac =$ ______________

16 $12efg \div (a \times a \times a) =$ ______________

ISBN: 9780170484084

Adding and subtracting

- Terms can only be added or subtracted if they are **'like' terms**.
- 'Like' terms must have exactly the same variables, and each variable must be raised to exactly the same power.
- Order does not matter.
- The sign does not matter.

Examples: The following **are** like terms: a^2b^5c, $2a^2b^5c$, $b^5a^2c^1$, $-a^2cb^5$

The following are **not** like terms (compared with the above):

a^2b^6c – the 'b' is raised to the power of 6, not 5

b^5c – there is no 'a' term

a^2b^5cd – includes a 'd' term

State whether each of the following pairs are like or unlike terms.

1 xy and yx ________________

2 $5x$ and $3x^2$ ________________

3 ab^2c and c^2ab ________________

4 x and $-2x$ ________________

5 xy^2 and $-0.1y^2x$ ________________

6 7 and 1 ________________

7 x^1 and x ________________

8 $4cd$ and $-9dce$ ________________

9 x^2yz and $-y^2zx$ ________________

10 $w^5x^3y^2z^1$ and $-y^2zw^5x^3$ ________________

- When adding or subtracting, you can combine only like terms.

Examples:

1 $\mathbf{7a + 2a} - a^2 = \mathbf{9a} - a^2$

2 $10 \mathbf{+ 5y} - 3 \mathbf{- y} = 7 \mathbf{+ 4y}$

3 $\mathbf{b^4 + 6b^4} - b^3 = \mathbf{7b^4} - b^3$

4 $2ab^2c \mathbf{- a^2bc - 2bca^2} = 2ab^2c \mathbf{- 3a^2bc}$

The **order** of the terms in your answer does not matter. → or $\mathbf{-3a^2bc} + 2ab^2c$

Simplify these by adding or subtracting like terms where possible.

11 $x + x + x + x =$ ________________

12 $5x^2 + x^2 + x =$ ________________

13 $12 + 3ab - 7 - ab =$ ________________

14 $7p^5q - pq^5 + 2p^5q + 4pq^5 =$ ________________

15 $2y^3 + 2y^2 + 6y^2 - 3y^3 =$ ________________

16 $x^2 + 5x + 7x + 35 =$ ________________

17 $11x + 6 - 4y - 5 =$ ________________

18 $6ab + 4a - 3b =$ ________________

19 $-5p + 17 - p + q =$ ________________

20 $x^2 - 9x - 3x + 27 =$ ________________

21 $y^2 - y + 8y - 8 =$ ________________

22 $3x^2 + 7xy - 8y^2 - xy + y^2 =$ ________________

23 $x^3 + x^2 - x^3 =$ ________________

24 $6x^2yz + 2xyz^2 - x^2yz - xyz^2 =$ ________________

ISBN: 9780170484084

Powers

The words **power**, **exponent** and **index** all mean the same thing.

coefficient — $\mathbf{5a^4}$ — power, exponent, index

base

Remember:

- $a^4 = a \times a \times a \times a$
- $a^1 = a$
- $a^0 = 1$

Multiplying terms that have powers

When multiplying terms that have powers, we **add** the indices.

Examples:

1 $a^5 \times a^2 = a^7$ — $5 + 2 = 7$

2 $\mathbf{3}a^4 \times \mathbf{5}a^9 = \mathbf{15}a^{13}$ — Always deal with the **coefficients** first.

3 $2ab \times 7ab^2 = 14a^2b^3$

Simplify these.

1 $p \times p \times p \times p \times p$

2 $a^4 \times a$

3 $b^7 \times b^2$

4 $g^3 \times g \times g^5$

5 $3a^2 \times 2a$

6 $2ab \times 5a^2b$

7 $15b^{10} \times 2b^5$

8 $fg^2 \times f \times g^4 \times fg$

9 $(-5y^6) \times (-3y^4)$

10 $(-a^2) \times (-b^3) \times a \times (-b)$

11 $-2y^4 \times 3y^2$

12 $2z^2 \times 4z^2 \times z^6$

ISBN: 9780170484084

Dividing terms that have powers

When dividing terms that have powers, we **subtract** the indices.

Examples: **1** $a^5 \div a^2 = a^3$ or $\frac{a^5}{a^2} = a^3$ 5 - 2 = 3

2 $\mathbf{12}a^6 \div \mathbf{6}a^2 = \mathbf{2}a^4$ Always deal with the **coefficients** first.

3 $\frac{14x^2}{7x^5} = \frac{2}{x^3}$

Simplify these.

1 $\frac{y^8}{y^2}$

____

2 $\frac{y^4}{y}$

____

3 $b^{12} \div b^3$

____

4 $\frac{10b^{10}}{2b^5}$

____

5 $\frac{15y^4}{30y}$

____

6 $\frac{x^2y^4}{y^3}$

____

7 $\frac{8y^4z}{2y^2}$

____

8 $\frac{-x^4}{2x^5}$

____

9 $\frac{12xy^2}{4y^7}$

____

10 $\frac{-2x^4}{6x^3}$

____

11 $\frac{-x^4y}{-3x^5y}$

____

12 $\frac{-10x^4y^2z^3}{-5x^7z}$

____

13 $3xy \div 18xy$

____

14 $\frac{7xyz^2}{14x^2yz^3}$

____

 ISBN: 9780170484084

Powers of powers

When finding a power of a power, we **multiply** the indices.

Examples:

1 $(a^2)^3 = a^2 \times a^2 \times a^2$
$= a^6$

2 × 3 = 6

2 $(4a^3)^2 = 4^2 \times (a^3)^2$
$= 16a^6$

or $(4a^3)^2 = 4a^3 \times 4a^3$
$= 16a^6$

3 $\left(\frac{1}{2}a^4\right)^3 = \left(\frac{1}{2}\right)^3 \times (a^4)^3$
$= \frac{1}{8}a^{12}$ or $\frac{a^{12}}{8}$ or $0.125a^{12}$

or $\left(\frac{1}{2}a^4\right)^3 = \left(\frac{1}{2}a^4\right) \times \left(\frac{1}{2}a^4\right) \times \left(\frac{1}{2}a^4\right)$
$= \frac{1}{8}a^{12}$ or $\frac{a^{12}}{8}$ or $0.125a^{12}$

Simplify these.

1 $(b^4)^2$

2 $(2x^3)^4$

3 $(-a^7)^2$

4 $(-y^3)^5$

5 $(-3x^4)^2$

6 $(-4x^4)^3$

7 $-2(5x^3)^2$

8 $(x^2y^4z^3)^3$

9 $-(-3a^3b^2)^2$

10 $2(-3xy^3)^2$

11 $\left(\frac{1}{2}x^2\right)^3$

12 $\left(-\frac{1}{5}x^4\right)^2$

ISBN: 9780170484084

Roots

- Finding a root is the opposite of finding a power.
- Roots are written as fractional powers.

Roots of numbers:

- $\sqrt{9}$ is the same as writing $\sqrt[2]{9}$ or $9^{\frac{1}{2}}$. $\sqrt{9}$ is the same as $\sqrt{3 \times 3} = \sqrt{3^2} = 3$.
- $\sqrt[3]{64}$ is the same as writing $64^{\frac{1}{3}}$. $\sqrt[3]{64}$ is the same as $\sqrt[3]{4 \times 4 \times 4} = \sqrt[3]{4^3} = 4$.

Roots of variables:

- $\sqrt{a^2}$ is the same as writing $\sqrt[2]{a^2}$ or $a^{\frac{2}{2}} = a$.
- $\sqrt[3]{a^3}$ is the same as writing $\sqrt[3]{a^3} = a^{\frac{3}{3}} = a$.

To find the **square** root of a power, you must **halve the power**: $\sqrt{a^{10}} = a^5$

$10 \div 2 = 5$

Examples:

1 $\sqrt{9x^8} = 3x^4$

2 $\sqrt{100x^6} = 10x^3$

Simplify these.

1 $\sqrt{25}$

______

2 $\sqrt{x^{10}}$

______

3 $\sqrt[3]{1000}$

______

4 $\sqrt{36x^6}$

______

5 $\sqrt{16x^{16}}$

______

6 $\sqrt[3]{64}$

______

7 $\sqrt{100x^{100}}$

______

8 $\sqrt{49x^8y^6}$

______

9 $\sqrt{(9x^{64})^2}$

______

10 $\sqrt{(2x^2y^4)^6}$

______

11 $\sqrt{25x^{10}y^6}$

______

12 $\sqrt{64x^{64}y^{16}z^{36}}$

______

ISBN: 9780170484084

Expanding and factorising

Expanding brackets

- In algebra, '**expand**' means **multiply** out all the brackets.
- After expanding, you are expected to collect the like terms in order to **simplify** the expression.

Examples: Expand and simplify the following.

1 $-2(5x + 3) = -10x - 6$

2 $4x(3 - 5x) = 12x - 20x^2$

3 $9 - 4(7x - 2) = 9 - 28x + 8$
$= -28x + 17$

Be careful!

4 $3(6x - 5) - (5x + 3) = 18x - 15 - 5x - 3$
$= 13x - 18$

Expand and simplify the following.

1 $5(3x - 4)$

2 $6x(2 - x)$

3 $-2x(3x - 5)$

4 $7(3x - 4) + 12$

5 $1 - 2(4 - 3x) - 5$

6 $10x - 3(7 - 2x)$

7 $4(3 - 2x) + 2(x - 5)$

8 $2(9x + 4) + 3(5 + 2x)$

9 $x(2x - 7) - (x - 4)$

10 $9(2x - 1) - 5(3 - x)$

11 $5x^2 - 2x(3 - 4x) + x^2$

12 $x(1 - 2x) - 7x(3x + 4)$

ISBN: 9780170484084

Factorising

- Factors are terms that are **multiplied** together (rather than added or subtracted). For example, 2 and 3 are factors of 6 because 2 x 3 = 6.
- Expressions with brackets are usually in factorised form.

Examples:

In each factorised expression there is an unwritten **x (times)** sign before the bracket.

Factorised form	Unfactorised (expanded) form
$2(3x + 4)$	$6x + 8$
$x(5 - 3x)$	$5x - 3x^2$
$3(2x^2 + x - 7)$	$6x^2 + 3x - 21$

- When factorising, you must factorise **completely**. There must be no common factor for the terms inside the brackets.
- You need to ask yourself: '**What is the biggest thing (or things) that will divide into every term?**'
- Do not use fractions when factorising.

The word '**completely**' or '**fully**' is understood, but not written here.

Examples: Factorise .. the following.

1 $12x - 18 = 6(2x - 3)$

Notice that $2(6x - 9)$ is not a fully factorised answer because 3 can divide into both $6x$ and 9.
Similarly, $3(4x - 6)$ is not a fully factorised answer because 2 divides into both $4x$ and 6.

2 $20x^2 + 30x = 10x(2x + 3)$

All of the following are true but are not fully factorised:

$2(10x^2 + 15x)$ $5(4x^2 + 6x)$ $10(2x^2 + 3x)$

$2x(10x + 15)$ $5x(4x + 6)$ $\frac{1}{2}(40x^2 + 60x)$

All of the brackets contain at least one common factor.

Circle the fully factorised answer for each of the following.

1 $16x + 8$

$2(8x + 4)$

$1(16x + 8)$

$8(2x + 1)$

$4(4x + 2)$

$\frac{1}{2}(32x + 16)$

2 $36x - 60$

$6(6x - 10)$

$1(36x - 60)$

$12(3x - 5)$

$3(12x - 20)$

$24(1.5x - 2.5)$

ISBN: 9780170484084

3 $100 + 20x$

$2(50 + 10x)$

$\frac{1}{2}(200 + 40x)$

$100\left(1 + \frac{1}{5}x\right)$

$20(5 + x)$

$10(10 + 2x)$

4 $60x - 24x^2$

$x(60 - 24x)$

$6(10x - 4x^2)$

$6x(10 - 4x)$

$12(5x - 2x^2)$

$5(3x - 4x^2)$

$12x(5 - 2x)$

5 $12xy^2 + 4xy$

$4xy(3y + 1)$

$4x(3y^2 + y)$

$2xy(6y + 2)$

$4y(3xy + x)$

$xy(12y + 4)$

6 $ab^4c^3 + a^2b^3c^4$

$ab^2c^3(b^2 + ab)$

$b^3c^3(ab + a^2c)$

$ab^3c^2(bc + ac^2)$

$ab^3c^3(b + ac)$

$ab^2c^2(b^2c + abc^2)$

Fully factorise the following.

7 $5x - 15$

____

8 $x^2 - 3x$

____

9 $8x + 16$

____

10 $24 - 18a$

____

11 $10y + 6$

____

12 $x^2 + 11x$

____

13 $2x + 3x^2$

____

14 $5x^2 + 15x$

____

15 $4y^2 + 12y$

____

16 $4a^2 + 2a$

____

17 $3x^2 + 15x$

____

18 $21y - 14y^3$

____

19 $10a^3 + 2a^2$

____

20 $12b^2 + 16b^3$

____

21 $xy + 9x^2y^2$

____

22 $14x^2y + 7xy^2$

____

23 $2ab^2c - a^2bc^3$

____

24 $24ab^2c^3 - 8a^2bc^3$

____

25 $10xy + 5x^2 - xy^2$

____

26 $12xy^2 + 4x^2y - 8x^2y^2$

____

Algebraic fractions

Simplifying fractions

- Some algebraic fractions will require simplifying.
- You may need to factorise before cancelling terms.

Examples:

1 $\frac{2x^2 + 4x}{3x^2y + 6xy} = \frac{2x(x+2)}{3xy(x+2)}$

(x + 2) is on the top and bottom of the fraction so it can be cancelled out.

$= \frac{2x}{3xy}$

The *x* can also be cancelled out.

$= \frac{2}{3y}$

2 $\frac{4xy + 8x^2}{2xy} = \frac{2xy + 4x^2}{xy}$

Divide each term by 2.

$= \frac{2y + 4x}{y}$

Each term has an *x* so this can also be cancelled out.

Simplify these fractions.

1 $\frac{16a + 18b}{2}$

2 $\frac{x - y}{7x - 7y}$

3 $\frac{a^2 - 6a}{a - 6}$

4 $\frac{21x^2 + 15x^3}{3x}$

5 $\frac{x + 4}{2x + 8}$

6 $\frac{(x + 2)(x - 1)}{3x + 6}$

7 $\frac{5xy}{20x + 10x^2}$

8 $\frac{7a^2 - 6a}{7a}$

9 $\frac{3ab + 4a^2}{5a}$

10 $\frac{18xy^2 + 12x^2y}{6x}$

11 $\frac{3bc^2}{21b^2c - 15bc}$

12 $\frac{8a + (2a)^2}{4a}$

ISBN: 9780170484084

Multiplying fractions

$$\frac{a}{b} \times \frac{c}{d} = \frac{ac}{bd}$$

Multiply the numerators and multiply the denominators.

Examples:

1 $\frac{a}{3} \times \frac{2}{5} = \frac{2a}{15}$

2 $\frac{2z}{7} \times \frac{z}{3} = \frac{2z^2}{21}$

3 $\frac{y}{2z} \times \frac{z}{5y} = \frac{1\cancel{yz}}{10\cancel{yz}} = \frac{1}{10}$

4 $\frac{2(y+1)}{4} \times \frac{3}{5} = \frac{6(y+1)}{20} = \frac{3(y+1)}{10}$

Try for yourself.

1 $\frac{3x}{y} \times \frac{6}{y}$

2 $\frac{7}{z} \times \frac{2}{z}$

3 $\frac{2z}{7} \times \frac{4}{5}$

4 $\frac{y}{2} \times \frac{4}{9}$

5 $\frac{5y}{6z} \times \frac{1}{y}$

6 $\frac{1}{4} \times \frac{2y}{3}$

7 $\frac{3}{4} \times \frac{y-5}{5}$

8 $\frac{2yz}{7} \times \frac{yz^2}{2}$

9 $\frac{2z}{3} \times \frac{4}{5z} \times \frac{z}{2}$

10 $\frac{3(x+1)}{2} \times \frac{x}{5}$

11 $\frac{6x^2yz}{5} \times \frac{yz}{3x}$

12 $\frac{3z}{y-1} \times \frac{y}{6z}$

ISBN: 9780170484084

Dividing fractions

$\div \rightarrow \times$

$$\frac{a}{b} \div \frac{c}{d} = \frac{a}{b} \times \frac{d}{c} = \frac{ad}{bc}$$

Change the ÷ sign to a **x** sign and flip the second fraction.

invert the second fraction → the reciprocal

Examples:

1 $\frac{z}{8} \div \frac{z}{2} = \frac{z}{8} \times \frac{2}{z}$

$= \frac{1}{4}$

2 $\frac{6y}{5} \div \frac{3}{y} = \frac{6y}{5} \times \frac{y}{3}$

$= \frac{2y^2}{5}$

Simplify these.

1 $\frac{12}{x} \div \frac{2}{x}$

2 $\frac{1}{x} \div \frac{1}{y}$

3 $\frac{3}{x} \div \frac{2}{x}$

4 $\frac{9y}{x} \div \frac{3}{x}$

5 $\frac{5x}{2} \div \frac{x}{10}$

6 $\frac{4x}{5} \div \frac{1}{15x}$

7 $\frac{x}{y^2} \div x$

8 $\frac{15y}{2x} \div \frac{3y}{x}$

9 $\frac{3x^2}{y} \div \frac{5x}{4y}$

10 $\frac{10x^2}{y^2} \div \frac{2x}{5y}$

 ISBN: 9780170484084

Adding and subtracting fractions

1 Where the denominators are the same

Adding or subtracting fractions with the **same denominator**:

$$\frac{a}{b}+\frac{c}{b}=\frac{a+c}{b} \text{ or } \frac{a}{b}-\frac{c}{b}=\frac{a-c}{b}$$

Examples:

1 $\frac{5x}{3}+\frac{2x}{3}=\frac{7x}{3}$

2 $\frac{4}{x}+\frac{3}{x}=\frac{7}{x}$

3 $\frac{x}{8}+\frac{3x}{8}=\frac{4x}{8}$

$=\frac{x}{2}$

4 $\frac{6x}{7}-\frac{x+2}{7}=\frac{6x-(x+2)}{7}$

$=\frac{5x-2}{7}$

Try for yourself.

1 $\frac{x}{9}+\frac{4x}{9}$

2 $\frac{7}{x}+\frac{2}{x}$

3 $\frac{5x}{7}-\frac{4x}{7}$

4 $\frac{7x}{y}-\frac{x}{y}$

5 $\frac{2x}{11}+\frac{4x}{11}-\frac{x}{11}$

6 $\frac{3x}{10}+\frac{2x}{10}$

7 $\frac{x}{5}+\frac{4x}{5}$

8 $\frac{7}{x}+\frac{4}{x}-\frac{5}{x}$

9 $\frac{2x-1}{7}+\frac{4x}{7}$

10 $\frac{5x+2}{9}+\frac{3(2x-1)}{9}$

11 $\frac{5(2x-3)}{12}-\frac{4x-3}{12}$

12 $\frac{6(3-2x)}{13}-\frac{x+5}{13}$

ISBN: 9780170484084

2 Where the denominators are different

Where fractions have **different denominators**, multiply both the numerator and the denominator by a number (or variable) that will create equal denominators.

$$\frac{a}{b}+\frac{c}{d}=\frac{a}{b}\times\frac{d}{d}+\frac{c}{d}\times\frac{b}{b}$$

$$=\frac{ad}{bd}+\frac{cb}{db}$$

$$=\frac{ad+bc}{bd}$$

$\frac{d}{d}=1$ and $\frac{b}{b}=1$

Examples:

1 $\frac{x}{8}+\frac{x}{3}=\frac{x}{8}\times\frac{3}{3}+\frac{x}{3}\times\frac{8}{8}=\frac{3x}{24}+\frac{8x}{24}$
$=\frac{11x}{24}$

2 $\frac{x}{2}+x=\frac{x}{2}+\frac{x}{1}\times\frac{2}{2}$
$=\frac{x}{2}+\frac{2x}{2}=\frac{3x}{2}$

3 $\frac{2x}{5}-\frac{1}{4}=\frac{2x}{5}\times\frac{4}{4}-\frac{1}{4}\times\frac{5}{5}=\frac{8x}{20}-\frac{5}{20}$
$=\frac{8x-5}{20}$

4 $\frac{2x+3}{2}-\frac{x}{3}=\frac{2x+3}{2}\times\frac{3}{3}-\frac{x}{3}\times\frac{2}{2}$
$=\frac{6x+9-2x}{6}=\frac{4x+9}{6}$

Simplify these.

1 $\frac{2x}{3}+\frac{x}{4}$

2 $\frac{3x}{7}+\frac{x}{2}$

3 $\frac{2x}{3}-\frac{x}{5}$

4 $\frac{x}{2}-\frac{3x}{7}$

5 $\frac{xy}{3}-\frac{1}{5}$

6 $4-\frac{x}{6}$

7 $\frac{y}{4}-\frac{x}{6}$

8 $3+\frac{x-2}{5}$

9 $\frac{2x-5}{5}+x$

10 $\frac{2x-1}{3}-\frac{1}{2}$

 ISBN: 9780170484084

3 Where the denominators are different algebraic expressions

Where fractions have different denominators which are **algebraic expressions**, use the same process.

Examples:

1 $$\frac{2}{x-3} + \frac{1}{x+4} = \frac{2}{x-3} \times \frac{\mathbf{x+4}}{\mathbf{x+4}} + \frac{1}{x+4} \times \frac{\mathbf{x-3}}{\mathbf{x-3}}$$

$$= \frac{2(x+4) + 1(x-3)}{(x-3)(x+4)}$$

$$= \frac{2x+8+x-3}{(x-3)(x+4)}$$

$$= \frac{3x+5}{(x-3)(x+4)}$$

2 $$\frac{x+1}{2x+5} - \frac{x-4}{3x-1} = \frac{x+1}{2x+5} \times \frac{\mathbf{3x-1}}{\mathbf{3x-1}} - \frac{x-4}{3x-1} \times \frac{\mathbf{2x+5}}{\mathbf{2x+5}}$$

$$= \frac{(x+1)(3x-1) - (x-4)(2x+5)}{(2x+5)(3x-1)}$$

$$= \frac{(3x^2 - x + 3x - 1) - (2x^2 + 5x - 8x - 20)}{(3x-1)(2x+5)}$$

$$= \frac{3x^2 + 2x - 1 - 2x^2 + 3x + 20}{(3x-1)(2x+5)}$$

$$= \frac{x^2 + 5x + 19}{(3x-1)(2x+5)}$$

Simplify these.

1 $\frac{2}{x+1} + \frac{3}{2x-4}$

2 $\frac{3}{3x-2} - \frac{x}{x+1}$

3 $\frac{x+3}{x+1} + \frac{x-1}{x+5}$

4 $\frac{x+1}{3x-2} + \frac{2x-3}{x+4}$

ISBN: 9780170484084

Substitution

- When substituting, you replace variables with numbers.
- It is important to remember **BEDMAS** when doing this.

Examples:

1 If $E = \frac{1}{2}mv^2$, $m = 20$ and $v = 12$, calculate the value of E.

$E = \frac{1}{2} \times 20 \times 12^2$

$= 1440$

Only the 12 is squared.

2 If $A = \frac{\sqrt{b+c}}{d}$, $b = 81$, $c = 19$ and $d = 2$, calculate the value of A.

$A = \frac{\sqrt{81+19}}{2}$

$= 5$

Don't forget to do the 81 + 19 **before** taking the square root.

If $b = 6$, $c = 3$, $d = -4$ and $e = -2$, calculate the values of A.

1 $A = (b - c)(c + d)$

2 $A = \frac{b}{2}(b - c)$

3 $A = \frac{1}{2}bd^2$

4 $A = \frac{b - d}{e}$

5 $A = \frac{\sqrt{bc + e}}{e}$

6 $A = b^2 - 5b - 7$

7 $A = d^2 - 7d - 13$

8 $A = \sqrt{b^2 - 4de}$

If $r = 3$ and $h = 4$, calculate the values of V. You may leave π in your answer.

9 $V = h^3 - r^3$

10 $V = \pi r^2 h$

11 $V = \frac{4}{3}\pi r^3$

12 $V = r^3 - \frac{1}{3}\pi r^3$

ISBN: 9780170484084

Solving linear equations

'Solve' means 'find a value for x'.

Rules: 1 You can do anything you like to an equation as long as you do the **same to both sides**.

2 There should be only **one equals sign** per line.

3 Collect all the variables on one side and numbers on the other side.

4 When you want to get rid of something, perform the **opposite** operation.

5 Your answer should always be in the form ***variable* = ...**

Trick: If you need to change the sign of everything, multiply **both** sides by **-1**.

One-step equations

Examples:

1 $y + 17.2 = 11$ **(- 17.2)**

$y = -6.2$

2 $1.23 - x = 4$ **(- 1.23)**

$-x = 2.77$ **(x -1)**

$x = -2.77$

3 $-6x = 20$ **(÷ -6)**

$x = -3.\dot{3}$

4 $\frac{x}{5} = -1.4$ **(x 5)**

$x = -7$

Solve the following.

1 $x - 0.6 = 1.1$

2 $x + 6.9 = 10$

3 $5x = 14$

4 $\frac{x}{8} = 0.25$

5 $x - 56 = -103$

6 $0.52 + x = -0.16$

7 $17.2 - x = 5.7$

8 $-1\frac{1}{2}x = -0.9$

9 $-x + 0.3 = 0.2$

10 $-143 - x = -10$

Two-step equations

Tricks: 1 Do the adding or subtracting **before** the multiplying or dividing.

2 If the variable (e.g. x) is on the right, **swap** the sides using the '=' sign as a centre.

e.g. $10.5 = 8x + 0.9$ → $8x + 0.9 = 10.5$

Examples:

1

$10.5 = 8x + 0.9$

$8x + 0.9 = 10.5$ **(- 0.9)**

$8x = 9.6$ **(÷ 8)**

$x = 1.2$

2

$5 - 2.5x = -12.5$ **(- 5)**

$-2.5x = -17.5$ **(÷ -2.5)**

$x = 7$

Solve the following.

1 $5x - 3 = 16$

2 $0.5x + 9 = 14$

3 $23 + 1.2x = 26.6$

4 $168 - 4x = 1$

5 $0.06x - 3.6 = -4.8$

6 $110x + 7.9 = 3.5$

7 $25x - 5 = 0$

8 $213 = 7x + 101$

9 $-5.4 = 0.2x - 14.8$

10 $-0.009 = 0.006 - 3x$

ISBN: 9780170484084

Equations with a variable on both sides

Remember – collect all the terms that include an 'x' on one side and the constants on the other.

Examples:

1

$1 - 18x = 100 + 4x$ **(− 4x)**

$1 - 22x = 100$ **(− 1)**

$-22x = 99$ **(÷ −22)**

$x = -4.5$

2

$9x + 1.1 = 2.7 - 7x$ **(+ 7x)**

$16x + 1.1 = 2.7$ **(− 1.1)**

$16x = 1.6$ **(÷ 16)**

$x = 0.1$

Solve the following.

1 $5x - 3 = 4x + 7$

2 $13 - 5x = 10x + 43$

3 $2x - 19 = 5x - 46$

4 $1 - 5x = x + 31$

5 $9x - 150 = 1.5x$

6 $21x - 18 = 8x + 21$

7 $1 - 27x = -39 - 19x$

8 $3 - x = 5x + 735$

9 $8x + 17 = 11x + 5$

10 $12x + 0.2 = -0.3 + 7x$

Equations with brackets

Expand the brackets, **then solve** as you did in the last exercise.

Examples:

1 $5(2x - 1) = 2(4 + x)$

$10x - 5 = 8 + 2x$ **(− 2x)**

$8x - 5 = 8$ **(+ 5)**

$8x = 13$ **(÷ 8)**

$x = 1.625$

2 $12(2 - x) = 3(5 - x) - 7(x + 1)$

$24 - 12x = 15 - 3x - 7x - 7$

$24 - 12x = 8 - 10x$ **(+ 10x)**

$24 - 2x = 8$ **(− 24)**

$-2x = -16$ **(÷ −2)**

$x = 8$

Solve the following.

1 $3(2x - 7) = 9(5 + x)$

2 $4(3 - x) = -2(5x + 3)$

3 $9(3x - 2) = -5(4 - 5x)$

4 $7(4 - 3x) = 3(x + 1) - 11$

5 $12(x - 3) = 5(1 - 7x) + 6$

6 $3(6x - 1) = 2(9 - x) - 1$

7 $7 - 5(3x - 1) = -3(5 + 2x)$

8 $3(2x - 5) = 5(7 - 3x) - (6 - x)$

ISBN: 9780170484084

Equations with fractions

Multiply **every term** by the lowest common multiple (LCM) of the denominators.

Multiply **every term** by 24 because it is LCM of 3 and 8.

Multiply **every term** by 6 because it is the LCM of 2 and 3.

Examples: 1

$$\frac{3x}{8} - 2 = \frac{x}{3}$$

$$\frac{3x}{8} \times \frac{\mathbf{24}}{\mathbf{1}} - 2 \times \mathbf{24} = \frac{x}{3} \times \frac{\mathbf{24}}{\mathbf{1}}$$

$$9x - 48 = 8x \qquad \mathbf{(+48)}$$

$$9x = 8x + 48 \qquad \mathbf{(-8x)}$$

$$x = 48$$

2

$$\frac{4 + 7x}{3} = \frac{5x + 1}{2}$$

$$\frac{4 + 7x}{3} \times \frac{\mathbf{6}}{\mathbf{1}} = \frac{5x + 1}{2} \times \frac{\mathbf{6}}{\mathbf{1}}$$

$$8 + 14x = 15x + 3 \qquad \mathbf{(-15x)}$$

$$8 - x = 3 \qquad \mathbf{(-8)}$$

$$-x = -5 \qquad \mathbf{(\times -1)}$$

$$x = 5$$

Solve the following.

1 $\frac{x}{5} + 2 = 9$

2 $\frac{x}{2} + \frac{2x}{5} = 9$

3 $\frac{x}{3} + \frac{x}{5} = 2$

4 $\frac{3x - 1}{2} = 7$

5 $\frac{9 - 5x}{3} = 4$

6 $\frac{3 - 4x}{4} = \frac{6x + 1}{5}$

7 $\frac{5 - 3x}{2} = \frac{7 - 2x}{5}$

8 $\frac{2x - 5}{3} - \frac{7 - x}{4} = 3$

ISBN: 9780170484084

Equations with powers

- Isolate the term with the variable on the left.
- Then write both sides so they are raised to the same power.
- Note that $\sqrt{9} = +3$. However, the expression $x^2 = 9$ has two solutions (+3 and −3) because both would satisfy the expression $x^2 = 9$.

Examples: Solve the following.

1
$x^3 = -125$
$x^3 = (-5)^3$ **($\sqrt[3]{\ }$)**
$x = -5$

2
$x^3 - 19 = 8$ **(+ 19)**
$x^3 = 27$
$x^3 = 3^3$
$x = 3$

3
$2x^4 = 32$ **(÷ 2)**
$x^4 = 16$
$x^4 = 2^4$ **($\sqrt[4]{\ }$)**
$x = \pm 2$

4
$\frac{1}{4}x^5 = -8$ **(x 4)**
$x^5 = -32$
$x^5 = (-2)^5$
$x = -2$

Solve the following.

1 $x^2 = 81$

2 $x^3 = 27$

3 $x^6 = 64$

4 $x^3 = -64$

5 $x^3 + 1 = 9$

6 $x^2 - 13 = 36$

7 $2x^3 = -54$

8 $3x^4 = 48$

9 $\frac{1}{2}x^2 = 18$

10 $\frac{1}{4}x^3 + 1 = -15$

ISBN: 9780170484084

Inequations

- Inequations have **<**, ≤, **>** or ≥ signs.
- You solve an inequation in **exactly** the same way as you solve an equation **except**:
  1. If you need to **multiply or divide** the equation by a **negative number**, you must **reverse the sign**.
  2. If you **swap sides** you must **reverse the sign**.

Examples:

1

$5x < 20$ **(÷ 5)**

$x < \frac{20}{5}$

$x < 4$

2

$4x + 1 \geq 17$ **(- 1)**

$4x \geq 16$ **(÷ 4)**

$x \geq \frac{16}{4}$

$x \geq 4$

3

$3 - 7x \leq 24$ **(- 3)**

$-7x \leq 21$ **(÷ -7)**

$x \geq -3$

4

$\frac{7 - x}{5} > \frac{2x - 1}{3}$ **(x 15)**

$21 - 3x > 10x - 5$ **(- 21)**

$-3x > 10x - 26$ **(- 10x)**

$-13x > -26$ **(÷ -13)**

$x < 2$

Dividing by a **negative** number ⇒ reverse the sign.

Solve the following.

1 $3x > 24$

2 $2x - 9 < 13$

3 $19 - 4x \leq 3$

4 $5x + 12 < 2x - 9$

5 $2(3 - 5x) < -4$

6 $3(2x - 5) \leq 7(4 + x)$

7 $1 - \frac{6x}{11} < 4$

8 $\frac{5 - 3x}{2} \geq \frac{7 - 2x}{5}$

Rearrangement of expressions

1 Where the subject appears once

- First put **the** term that contains the required subject on the left.
- Get rid of fractions by multiplying by the LCM of the denominator.
- Get rid of square roots, isolating the square root on one side and then squaring both sides.
- Use normal equation-solving rules to isolate the subject.

Examples:

1 Make r the subject of $V = \pi r^2 h$.

$\pi r^2 h = V$ **(÷ πh)**

$r^2 = \frac{V}{\pi h}$ **(√ both sides)**

$r = \sqrt{\frac{V}{\pi h}}$

2 Make w the subject of $A = \pi\sqrt{\frac{w}{g}}$.

$\pi\sqrt{\frac{w}{g}} = A$ **(÷ π)**

$\sqrt{\frac{w}{g}} = \frac{A}{\pi}$ **(square)**

Get the √ on its own on the left.

$\frac{w}{g} = \frac{A^2}{\pi^2}$ **(x g)**

$w = \frac{A^2 g}{\pi^2}$

3 Rewrite the formula $A = \frac{(a+b)}{2}h$ with a as the subject.

$\frac{(a+b)}{2}h = A$ **(x 2)**

$(a+b)h = 2A$ **(expand brackets)**

$ah + bh = 2A$ **(- bh)**

$ah = 2A - bh$ **(÷ h)**

$a = \frac{2A - bh}{h}$

4 Express a in terms of b if $(a^4)^3 = (b^3)^2$.

$(a^4)^3 = (b^3)^2$

$a^{12} = b^6$

$a^{12} = (b^{\frac{1}{2}})^{12}$

$a = b^{\frac{1}{2}}$ or $\sqrt{b}$

Rearrange the following equations.

1 Make r the subject of $A = \pi r^2$.

____

____

____

____

2 Make P the subject of $I = \frac{PRT}{100}$.

____

____

____

____

3 Rearrange $y = mx + c$ so that x is the subject.

____

____

____

____

4 Make x the subject of $y - 7 = 5x + 1$.

____

____

____

____

ISBN: 9780170484084

5 Rewrite the expression $y = ax^2 + b$ with x as the subject.

6 Make x the subject of $3y - 7 = 2x + 5$.

7 Make x the subject of $y - 1 = 3(2x + 8)$.

8 Make r the subject of $V = \frac{1}{3}r^2h$.

9 Make x the subject of $y - 5 = \frac{2x - 1}{6}$.

10 Make x the subject of $y + 6 = \frac{3}{4}(x - 2)$.

11 Express a in terms of b if $a^2 \times a^4 = (b^3)^2$.

12 Rewrite the formula $A = \frac{(a + b)h}{2}$ with h as the subject.

13 Make p the subject of $3t(p + q) = a$.

14 The formula for converting degrees Fahrenheit into degrees Celsius is $F = 32 + \frac{9C}{5}$. Make C the subject.

2 Where the subject appears twice

- If the subject appears in several terms, collect all of these to one side and factorise.

Examples:

1 Make a the subject of $5ab - 3 = 2a$.

$5ab - 3 = 2a$ **(− 2a)**

$5ab - 2a = 3$ **(factorise)**

$a(5b - 2) = 3$ **(÷ (5b − 2))**

$a = \frac{3}{5b - 2}$

Both terms with **a** in them on the left.

2 Make a the subject of $\frac{1}{b} = \frac{2}{a} + \frac{3}{c}$.

$\frac{1}{b} = \frac{2}{a} + \frac{3}{c}$ **(x $\frac{abc}{1}$)**

$\frac{1}{b} \times \frac{abc}{1} = \frac{2}{a} \times \frac{abc}{1} + \frac{3}{c} \times \frac{abc}{1}$

$ac = 2bc + 3ab$ **(− 3ab)**

$ac - 3ab = 2bc$ **(factorise)**

$a(c - 3b) = 2bc$ **(÷ (c − 3b))**

$a = \frac{2bc}{c - 3b}$

Make a the subject of each of the following expressions.

1 $2a = 6 - ab$

2 $5a = 3b - ab$

3 $2(a - 1) = ab$

4 $2(3a - b) = 5b(1 - a)$

5 $b + 1 = \frac{4}{a}$

6 $\frac{3}{a} = 2 + \frac{1}{b}$

 ISBN: 9780170484084

Quadratic expressions

- An expression in which the **highest power** of the variable is **2** is known as a quadratic expression.
- Usually these contain an $\mathbf{x^2}$, but in factorised form quadratic expressions may look like $(x \pm a)(x \pm b)$ or $x(x \pm a)$.
- When graphed, quadratics form curves known as **parabolas**:

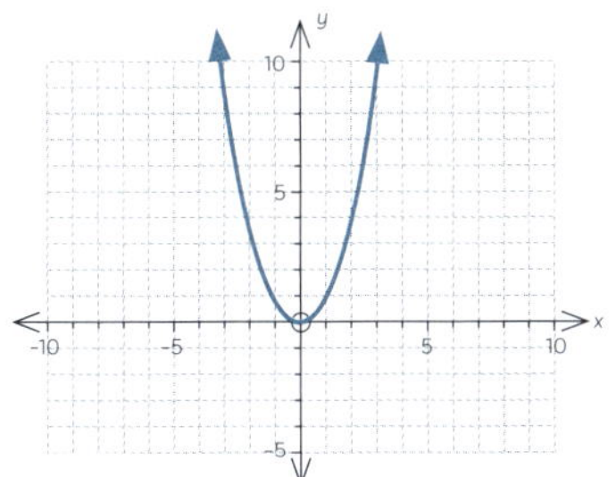

or

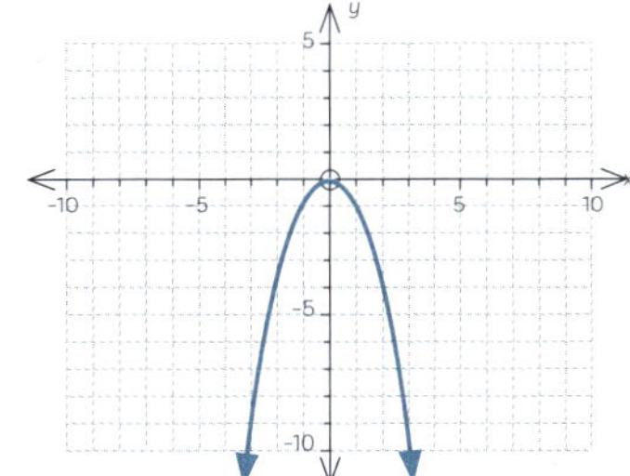

Expanding quadratic expressions

Remember **FOIL**: Multiply the **F**irsts
Outers
Inners
Lasts

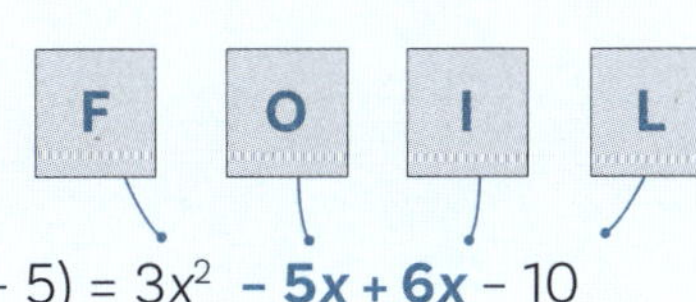

e.g. $(x + 2)(3x - 5) = 3x^2 - 5x + 6x - 10$
$= 3x^2 + x - 10$

Combine like terms

Answers are usually written in this order: x^2 **term, x term, number**

Examples:

1 $(x + 4)(x + 5) = x^2 + 5x + 4x + 20$
$= x^2 + 9x + 20$

2 $(5x + 3)(x - 2) = 5x^2 - 10x + 3x - 6$
$= 5x^2 - 7x - 6$

3 $(x + 7)^2 = (x + 7)(x + 7)$
$= x^2 + 7x + 7x + 49$
$= x^2 + 14x + 49$

4 $(4x - 1)^2 = (4x - 1)(4x - 1)$
$= 16x^2 - 4x - 4x + 1$
$= 16x^2 - 8x + 1$

5 $(x + \mathbf{4})(x - \mathbf{4}) = x^2 - 4x + 4x - 16$
$= x^2 - 16$

Expand and simplify these.

1 $(x + 3)(x + 4)$

____

____

2 $(x + 7)(x - 2)$

____

____

ISBN: 9780170484084

3 $(x + 3)(x - 9)$

4 $(x - 3)(x - 10)$

5 $(5 + x)(x + 8)$

6 $(1 + 3x)(x - 7)$

7 $(x + 8)^2$

8 $(x - 5)^2$

9 $(3x + 1)(x - 2)$

10 $(3x - 1)(2x + 6)$

11 $(3 - x)^2$

12 $(3x - 5)^2$

13 $3x(2x - 5)$

14 $(x + 1)(x - 1)$

15 $(5x - 2)(4x + 3)$

16 $(7 - 2x)(4 + 3x)$

17 $(4x - 3)(4x + 3)$

18 $(5 - 2x)(5 + 2x)$

ISBN: 9780170484084

Factorising quadratic expressions

Steps:

1 List all the factors of the constant: $x^2 + 3x - 40$

1, 40
2, 20
4, 10
5, 8

2 Select the pair that could add or subtract to give the coefficient of x: $x^2 + 3x - 40$

$-5 + 8 = +3$

3 The factors are $(x - 5)(x + 8)$

4 **Check** your answer by expanding the brackets using **FOIL** – you should get the original expression: $(x - 5)(x + 8) = x^2 + 3x - 40$

Examples:

1 $x^2 + 7x + 12 = (x + 3)(x + 4)$

1, 12
2, 6
3, 4

2 $x^2 - 11x + 30 = (x - 5)(x - 6)$

1, 30
2, 15
3, 10
-5, -6

3 $x^2 - 7x - 18 = (x + 2)(x - 9)$

1, 18
+2, -9
3, 6

4 $x^2 - 36 = x^2 + 0x - 36 = (x + 6)(x - 6)$

Notice the squares.

Insert a 'fake' x term.

1, 36
2, 18
3, 12
4, 9
+6, -6

Note: Any quadratic expression in the form $a^2 - b^2$ will factorise to $(a + b)(a - b)$. This is often called the 'difference of two squares'.

Factorise the following.

1 $x^2 + 6x + 8$

____

____

2 $x^2 + 8x + 7$

____

____

3 $x^2 + 13x + 36$

4 $x^2 + 4x - 12$

5 $x^2 + 19x + 60$

6 $x^2 - 9x + 18$

7 $x^2 - 2x - 24$

8 $x^2 - 10x + 25$

9 $x^2 - 5x - 6$

10 $x^2 - x - 6$

11 $x^2 - 5x + 6$

12 $x^2 + 5x - 6$

13 $x^2 - 25$

14 $49 - x^2$

15 $9 - 16x^2$

16 $25 - 9x^2$

17 $24 + 11x + x^2$

18 $15 - 8x + x^2$

ISBN: 9780170484084

Straight lines

Coordinates

- A positive **x** coordinate tells you how far to move to the **right**.
- A positive **y** coordinate tells you how far to move **up**.
- Coordinates are written in brackets and are in **alphabetical order**: **(x, y)**.
- We use the word 'ax**i**s' for one axis, and the word 'ax**e**s' for more than one.

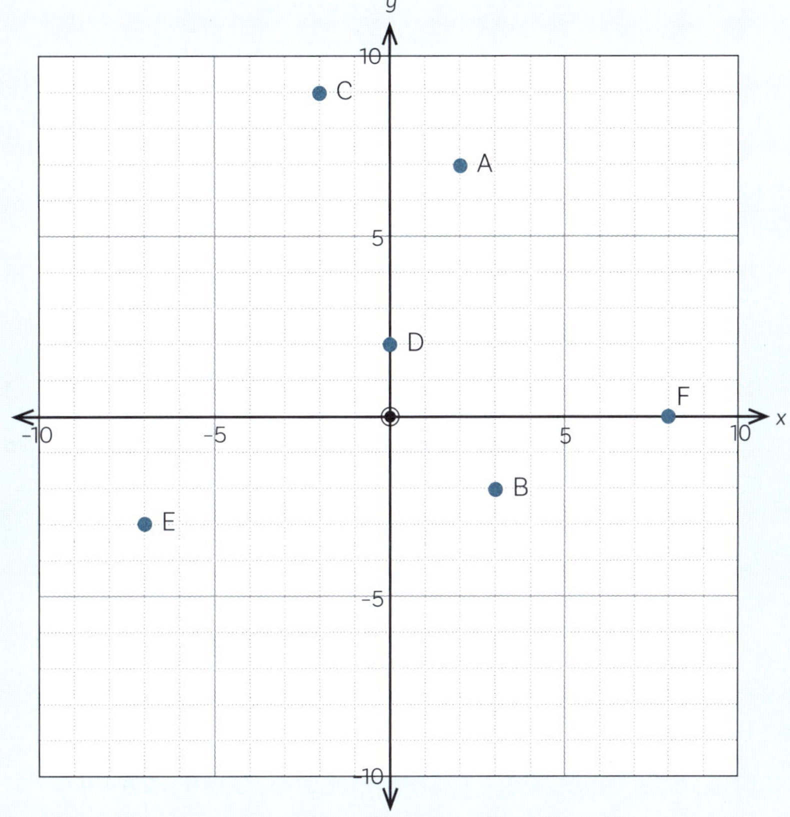

1 Write down the coordinates of the lettered points shown below.

A	(,)	B	(,)
C	(,)	D	(,)
E	(,)	F	(,)

2 Plot these points on the axis.

G	(-5, 4)	H	(1, -6)
I	(-3, 0)	J	(4, 2)
K	(0, 6)	L	(-4, -8)

Linear patterns with discrete data

- **Discrete** data is data that can be **counted**, e.g. 'number of ...', prices of items that can be bought with only \$5 notes, heights that are measured to the nearest centimetre.
- You need to be able to continue a pattern, plot these on a graph, find a rule and use it.

Example: Alex makes patterns with buttons as shown.

Pattern 1 Pattern 2 Pattern 3

a Complete the table.

Pattern # (n)	# of buttons (B)
1	6
2	8
3	10
4	**12**
5	**14**
6	**16**

(+2 between each row)

Look for the pattern: in this case, **+2**.

b Plot the points on the graph.

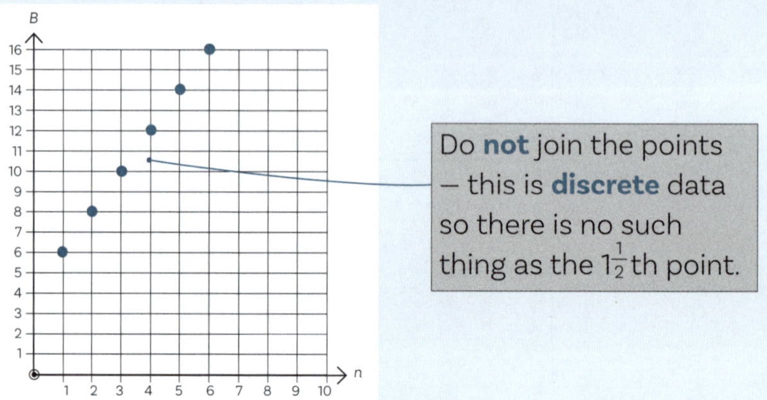

c Give the equation for the number of buttons (B) in any pattern (n).

Use the format $B =$ **+2** $n +$ **+4** So the equation is **$B = 2n + 4$**

Insert the **+2** from the table.

Insert the # of buttons for pattern number 0.

d Use your equation to find how many buttons would be in his 20th pattern.

20th pattern $\Rightarrow n = 20$ ∴ **$B = 2$** x 20 **+ 4** = 44. So the 20th pattern would have 44 buttons.

e Explain how the equation relates to the pattern.

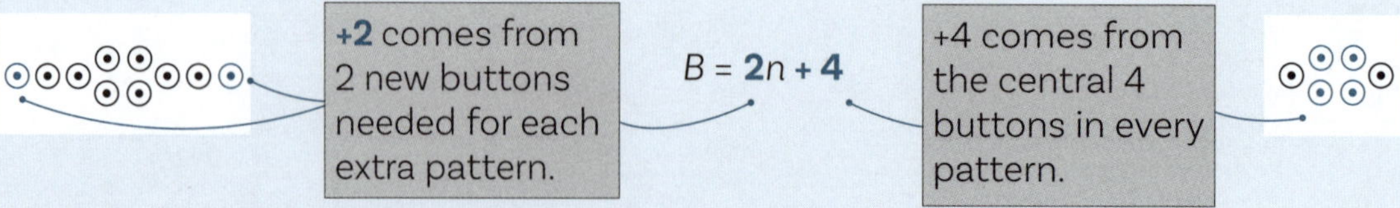

f Which pattern would have 92 buttons?

$92 = 2n + 4$ **(− 4)**

$88 = 2n$ **(÷ 2)**

∴ $n = 44$. So pattern 44 has 92 buttons.

ISBN: 9780170484084

Answer the following questions.

1 Holly makes the following pattern with buttons.

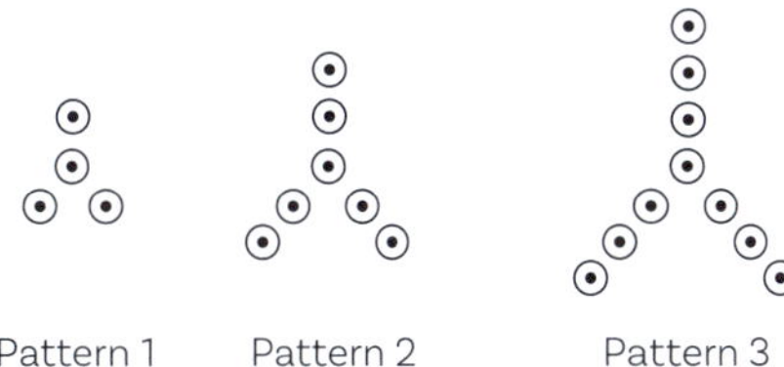

a Complete the table, and use it to find the equation.

Pattern # **(n)**	**# of buttons (B)**
1	4
2	7
3	10
4	
5	
6	

Equation: B = _______ n + _______

b Plot the points on the graph.

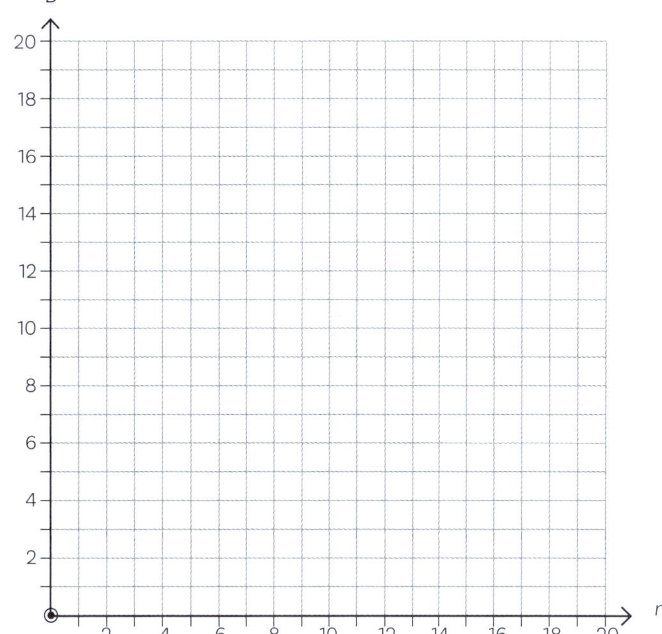

c How many buttons are needed for the 30th pattern?

30th pattern $\Rightarrow$ $n = 30 \therefore B$ = _______ x 30 + _______ = _______

So the 30th pattern needs _______ buttons.

d Which pattern would have 271 buttons?

271 buttons $\Rightarrow$ 271 = _______ n + _______

= _______________

So the _______ pattern would need 271 buttons.

e How does the equation relate to the pattern?

B = _______ n + _______

_______________________ _______________________

_______________________ _______________________

2 Holly makes another pattern with black and white buttons.

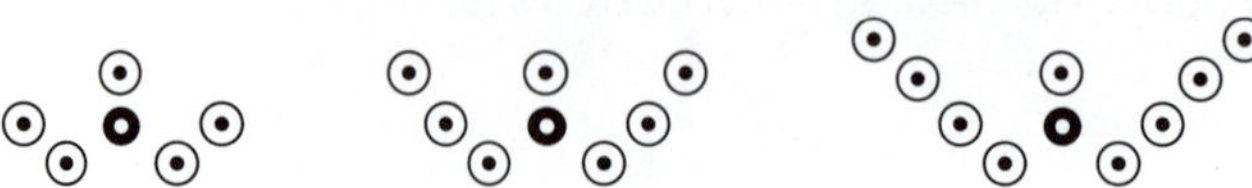

Pattern 1 Pattern 2 Pattern 3

a Complete the table, and use it to find the equation.

Pattern # (n)	# of buttons (B)
1	6
2	8
3	10
4	
5	
6	

Equation: B = ______ n + ______

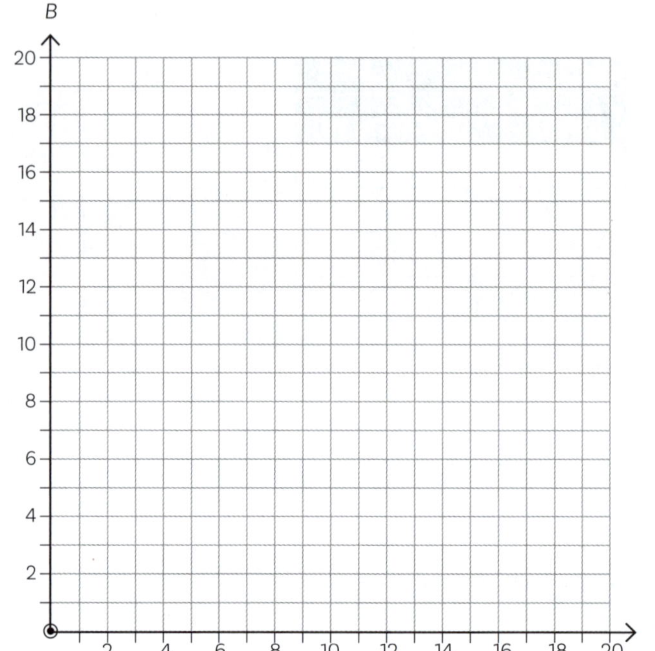

b Plot the points on the graph.

c How many buttons are needed for the 30th pattern?

30th pattern ⇒ $n = 30 \therefore B$ = ______ x 30 + ______ = ______

So the 30th pattern needs ______ buttons.

d Which pattern would have 164 buttons?

164 buttons ⇒ 164 = ______ n + ______

= ______________

So the ______ pattern would need 164 buttons.

e How does the equation relate to the pattern?

B = ______ n + ______

______________ ______________

______________ ______________

ISBN: 9780170484084

3 Nick makes the following pattern with matchsticks. How many matchsticks are needed for the 25th pattern?

Pattern 1

Pattern 2

Pattern 3

a Complete the table, and use it to find the equation.

Pattern # (n)	# of matchsticks (M)
1	5
2	9
3	13
4	
5	
6	

Equation: M = ______ n + ______

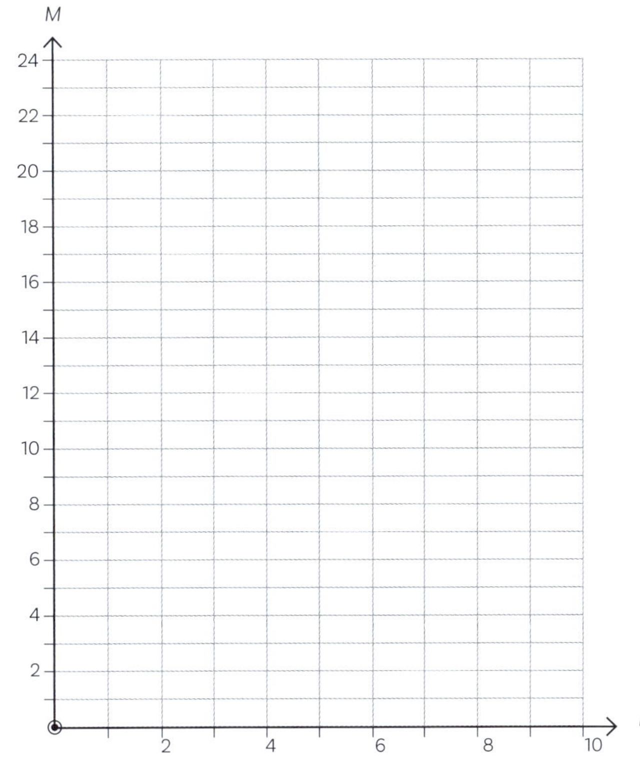

b Plot the points on the graph.

c How many matchsticks are needed for the 30th pattern?

30th pattern $\Rightarrow n = 30 \therefore M$ = ______ x 30 + ______ = ______

So the 30th pattern needs ______ matchsticks.

d Which pattern would have 245 matchsticks?

245 matchsticks $\Rightarrow$ 245 = ______ n + ______

= ______________

So the ______ pattern would need 245 matchsticks.

e How does the equation relate to the pattern?

M = ______ n + ______

______________ ______________

______________ ______________

ISBN: 9780170484084

4 Tahu's dad will give him $40 if he passes NCEA Level 1, plus $10 for each standard in which he earns Merit or Excellence. The table below shows how much Tahu will get for his first few Merit or Excellence standards, assuming he passes Level 1.

a Complete the table, and use it to find the equation.

# of Merit or Excellence standards (s)	$ earned (D)
1	50
2	
3	
4	
5	
6	

Equation: D = ______ s + ______

b Plot the points on the graph.

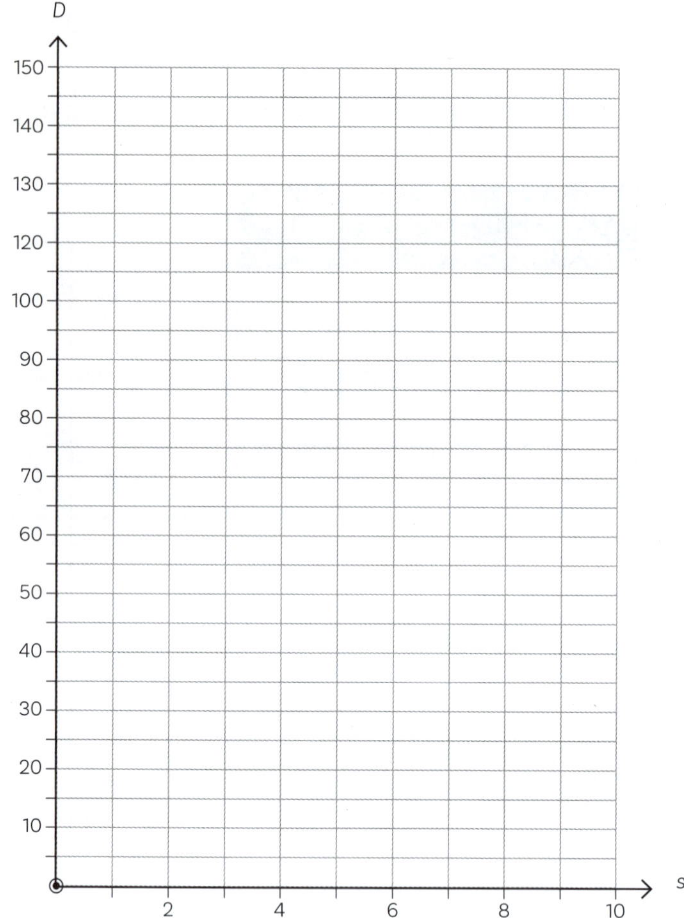

c How much does he earn if he passes 13 standards with Merit or Excellence?

13th pattern ⇒ n = 13 ∴ D = ______ x 13 + ______ = ______

So if he passes 13 standards with Merit or Excellence, he earns $ ______

d If he earns $210, how many standards did he pass with Merit or Excellence?

210 standards ⇒ 210 = ______ s + ______

= ______________

So if he earned $210, he passed ______ standards with Merit or Excellence.

e How does the equation relate to the pattern?

D = ______ s + ______

______________________ ______________________

______________________ ______________________

 ISBN: 9780170484084

The gradient of a line

The gradient is the **steepness**, or **slope**, of a line.

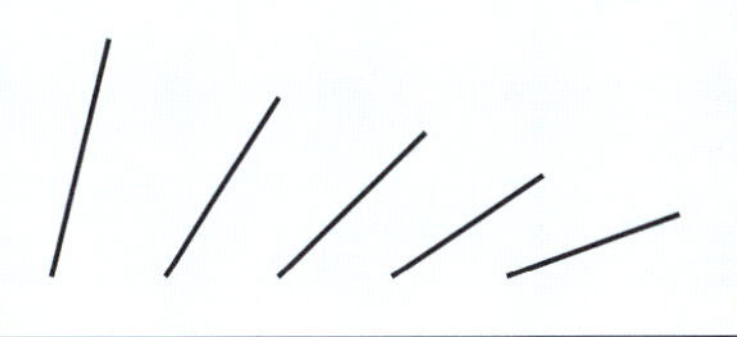

These lines all have **positive** gradients.

These lines all have **negative** gradients.

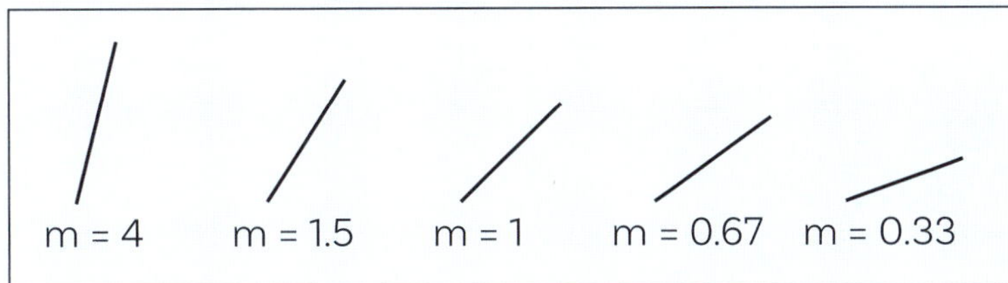

The **steeper** the line, the **bigger** the gradient.

The gradient is calculated using the formula $\boldsymbol{m} = \dfrac{\textbf{change in } \boldsymbol{y}}{\textbf{change in } \boldsymbol{x}}$ **or** $\dfrac{\textbf{rise}}{\textbf{run}}$.

The easiest way to do this is to draw a right-angled triangle on the line.

$$m = \frac{\text{rise}}{\text{run}} = \frac{5}{6}$$

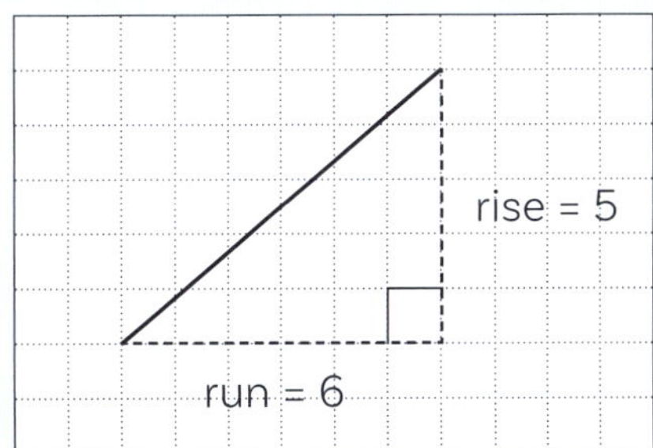

This time the gradient is *negative*.

$$m = \frac{\text{rise}}{\text{run}} = -\frac{3}{7}$$

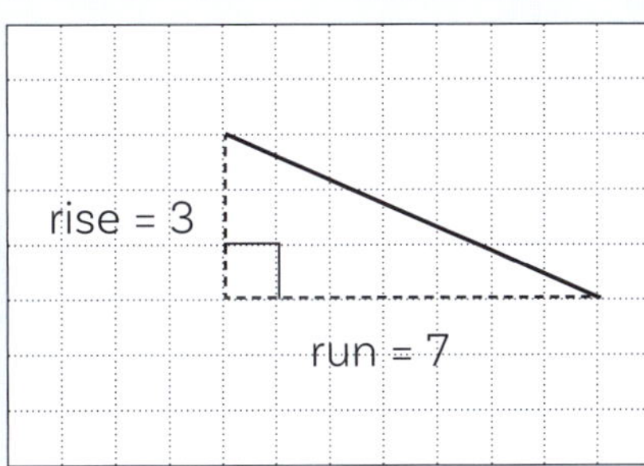

ISBN: 9780170484084

Horizontal lines

$$m = \frac{\text{rise}}{\text{run}}$$
$$= \frac{0}{7}$$
$$= 0$$

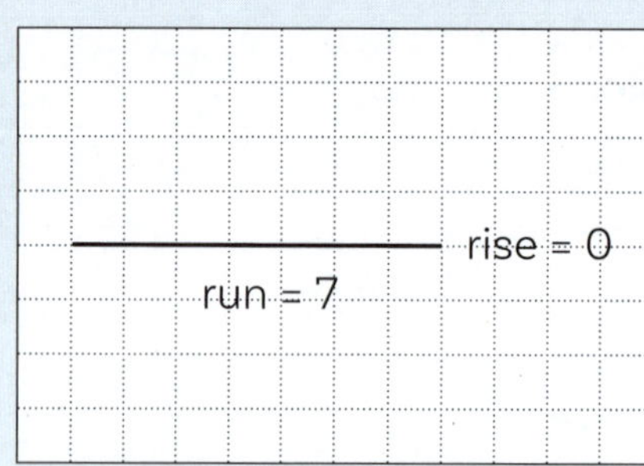

Vertical lines

$$m = \frac{\text{rise}}{\text{run}}$$
$$= \frac{5}{0}$$
$$= \text{undefined}$$

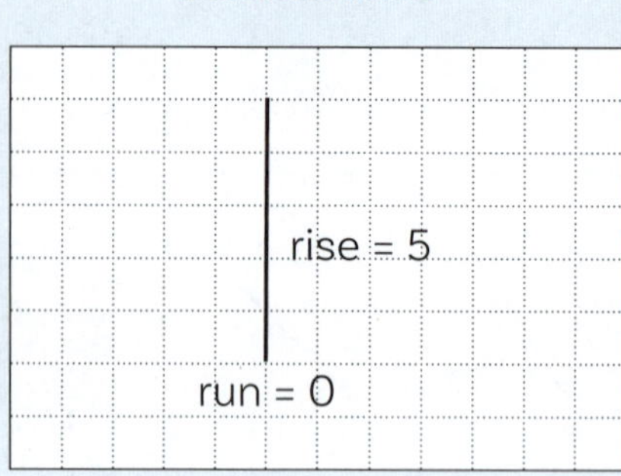

1 Calculate the gradients of these lines.

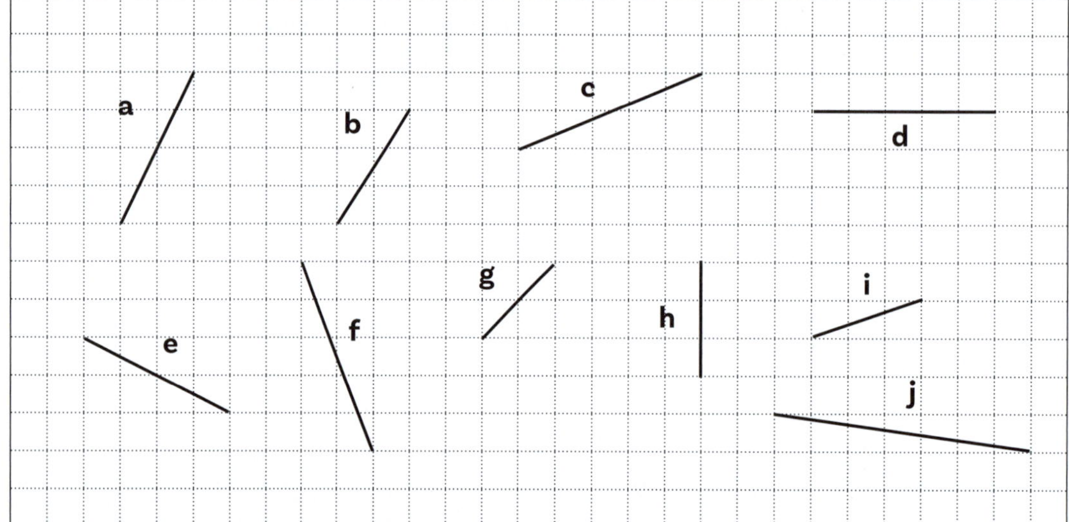

Gradient **a** = $\frac{\text{rise}}{\text{run}}$ = ______________________

Gradient **b** = ______________________

Gradient **c** = ______________________

Gradient **d** = ______________________

Gradient **e** = ______________________

Gradient **f** = ______________________

Gradient **g** = ______________________

Gradient **h** = ______________________

Gradient **i** = ______________________

Gradient **j** = ______________________

ISBN: 9780170484084

2 Draw line segments to show these gradients.

a $m = \frac{1}{3}$

b $m = 3$

c $m = \frac{3}{5}$

d $m = -2$

e $m = \frac{4}{7}$

f $m = -\frac{1}{4}$

g $m = -4$

h $m = -1\frac{1}{3}$

i $m = 0$

j $m = -1$

Gradients with different scales

- In practical situations, the scales on the axes are usually different.
- You need to be very careful when drawing or calculating these.

1 Drawing gradients

Example 1: m = 60, starting from the point (0, 100)

Step 1: Write the gradient as equivalent fractions:

$$m = \frac{\text{rise}}{\text{run}} = \frac{60}{1} = \frac{120}{2} = \mathbf{\frac{300}{5}} = \frac{600}{10} \text{ etc.}$$

Step 2: Select one that works with the given scales, and use it to draw another point. Join the points.

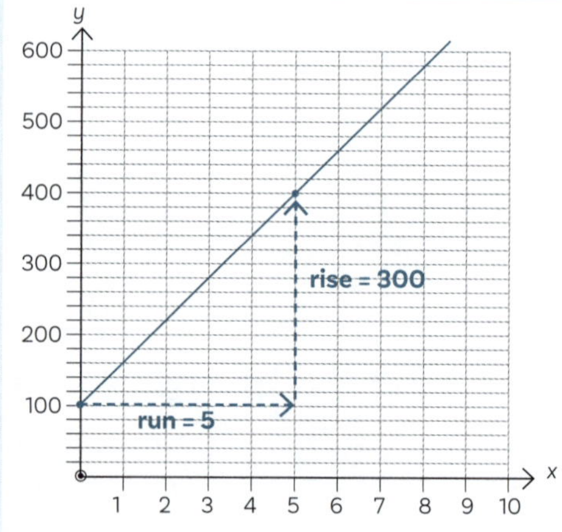

Example 2: $m = -\frac{2}{5}$, starting from the point (0, 28)

Step 1: Write the gradient as equivalent fractions:

$$m = \frac{\text{rise}}{\text{run}} = -\frac{2}{5} = -\frac{4}{10} = \mathbf{-\frac{20}{50}} \text{ etc.}$$

Step 2: Select one that works with the given scales, and use it to draw another point. Join the points.

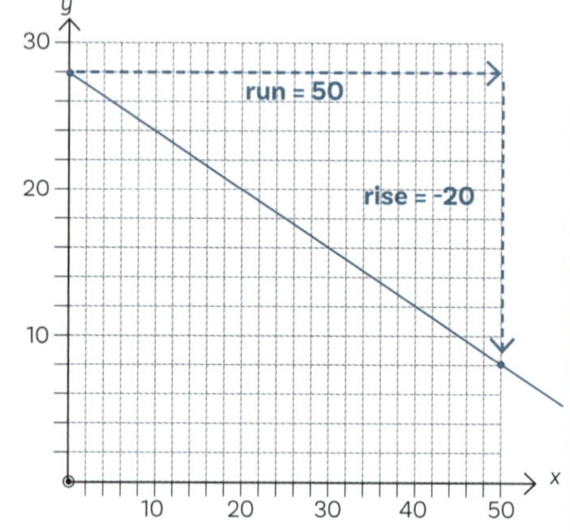

Draw the following gradients.

1 m = 40, starting from the point (0, 200)

$m = \frac{40}{1} =$

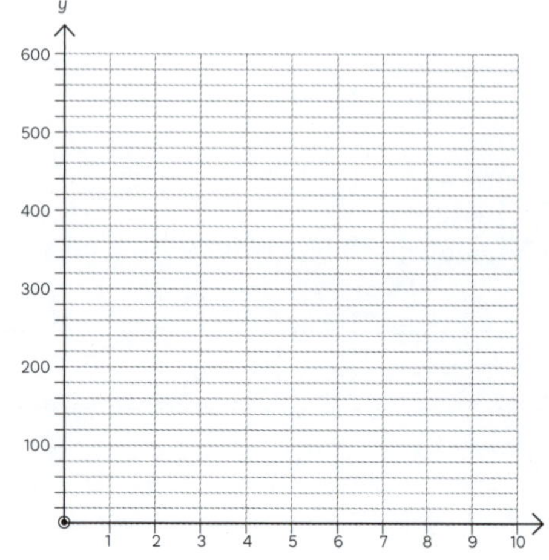

2 $m = -\frac{3}{5}$, starting from the point (0, 30)

$m = -\frac{3}{5} =$

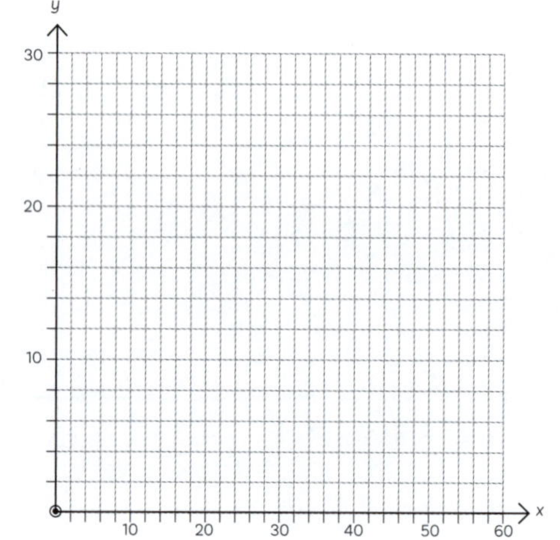

 ISBN: 9780170484084

3 $m = 3$, starting from the point (0, 10)

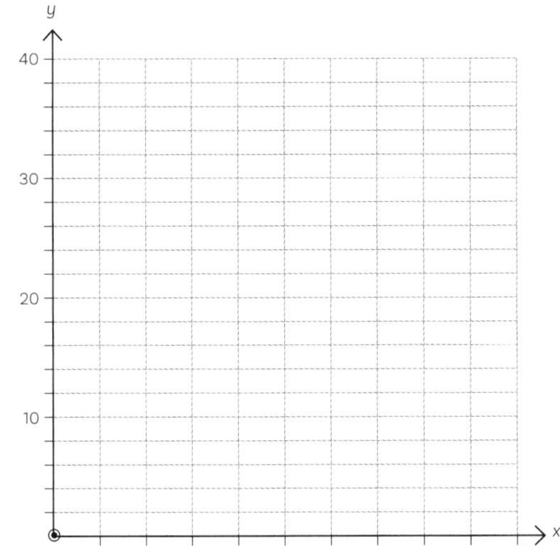

4 $m = -\frac{1}{2}$, starting from the point (0, 20)

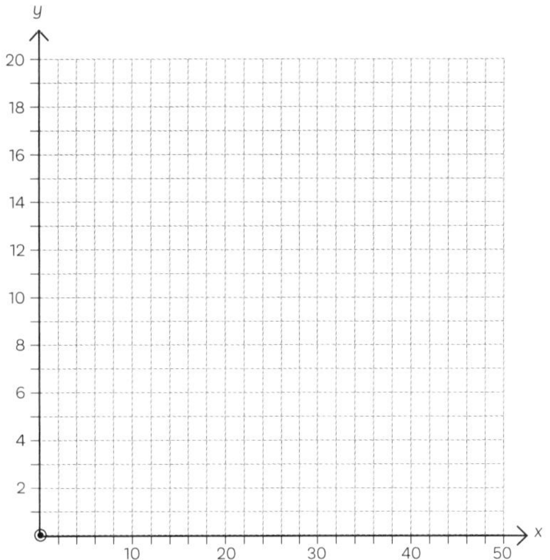

5 $m = -6$, starting from the point (0, 90)

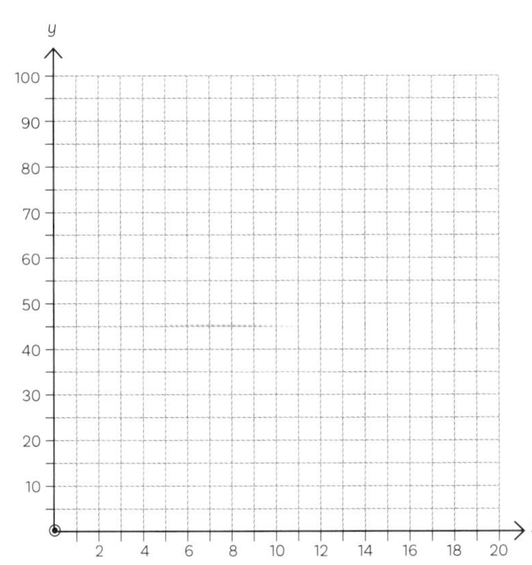

6 $m = \frac{1}{3}$, starting from the point (0, 4)

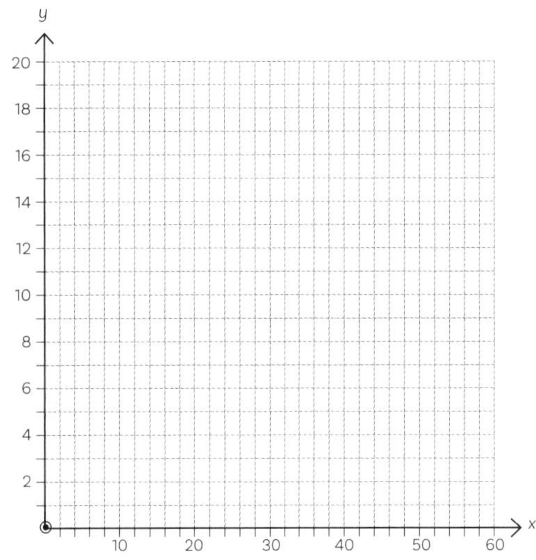

7 $m = -2$, starting from the point (0, 26)

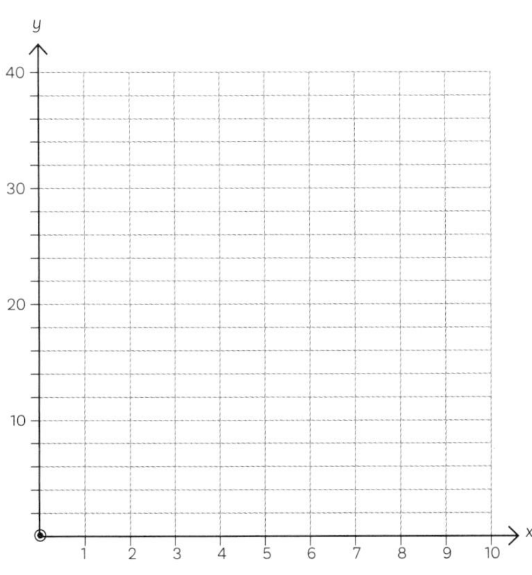

8 $m = 12$, starting from the point (0, 8)

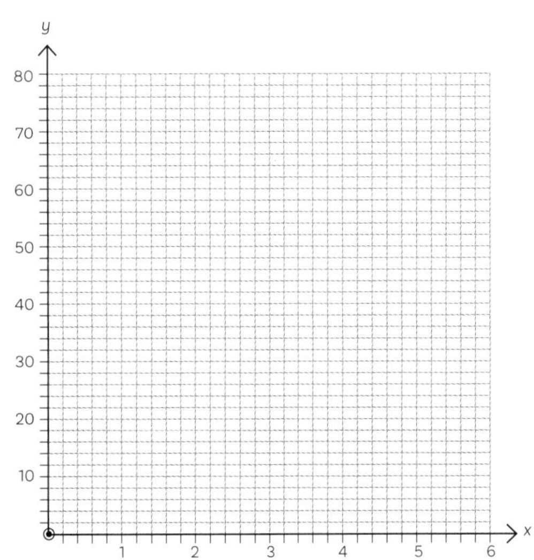

ISBN: 9780170484084

2 Calculating gradients

Example 1:

Step 1: Mark two points that the line clearly passes through and draw a right-angled triangle.

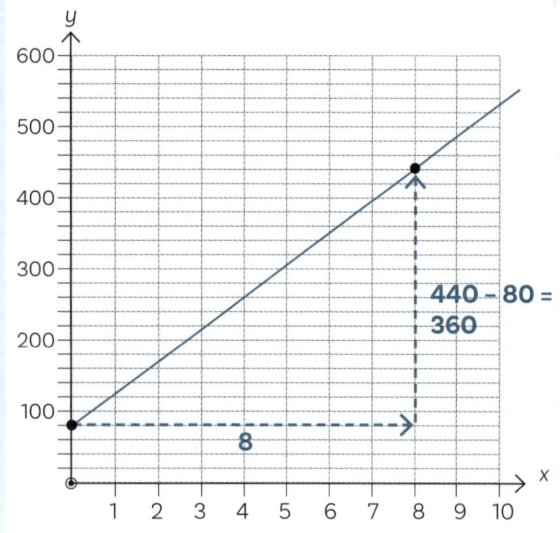

Step 2: Substitute:

$$m = \frac{\text{rise}}{\text{run}} = \frac{360}{8} = 45$$

Example 2:

Step 1: Mark two points that the line clearly passes through and draw a right-angled triangle.

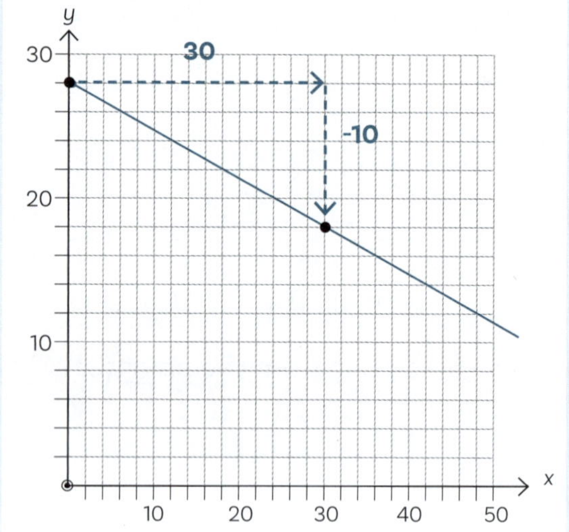

Step 2: Substitute:

$$m = \frac{\text{rise}}{\text{run}} = -\frac{10}{30} = -\frac{1}{3}$$

Calculate the following gradients.

1

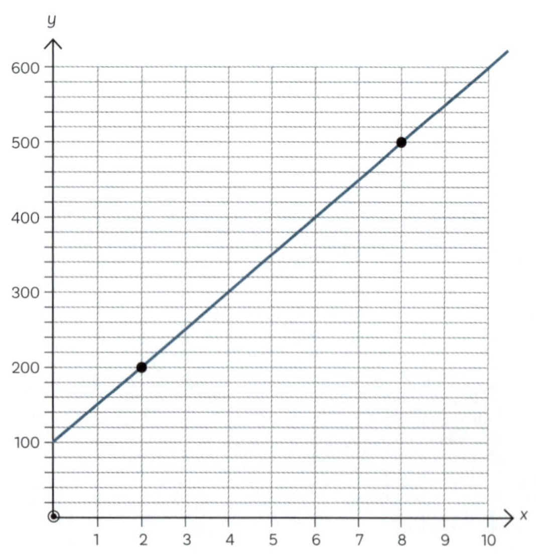

$$m = \frac{\text{rise}}{\text{run}} = \frac{\quad}{\quad} =$$

2

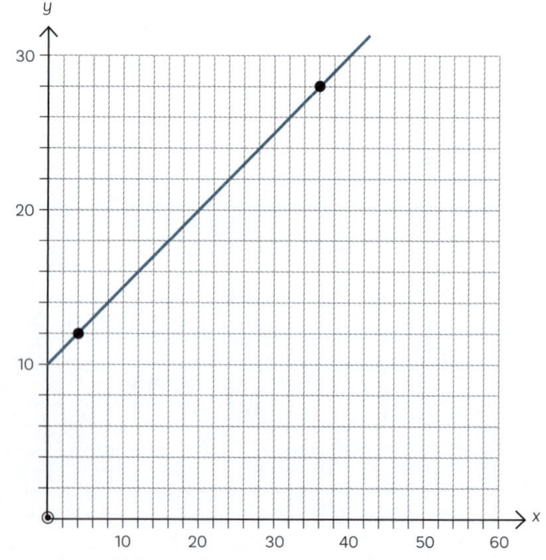

$$m = \frac{\text{rise}}{\text{run}} = \frac{\quad}{\quad} =$$

 ISBN: 9780170484084

3

y
20
18
16
14
12
10
8
6
4
2
10
20
30
40
50
x

4

y
40
30
20
10
1
2
3
4
5
6
7
8
9
10
x

5

y
70
60
50
40
30
20
10
2
4
6
8
10
12
14
16
18
20
x

6

y
20
18
16
14
12
10
8
6
4
2
10
20
30
40
50
60
x

7

y
40
30
20
10
10
20
30
40
50
60
70
80
90
100
x

8

y
80
70
60
50
40
30
20
10
1
2
3
4
5
6
x

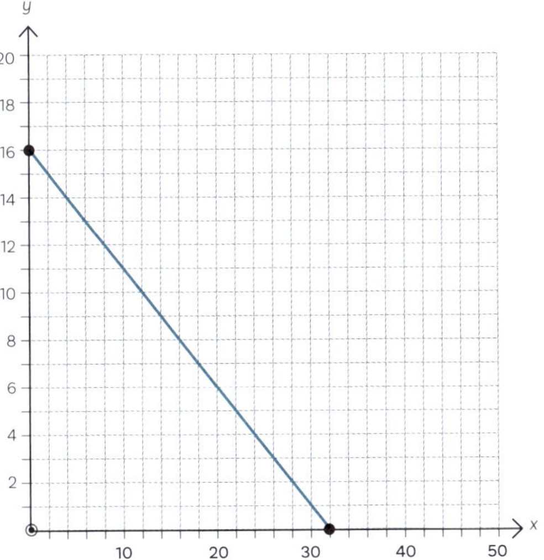

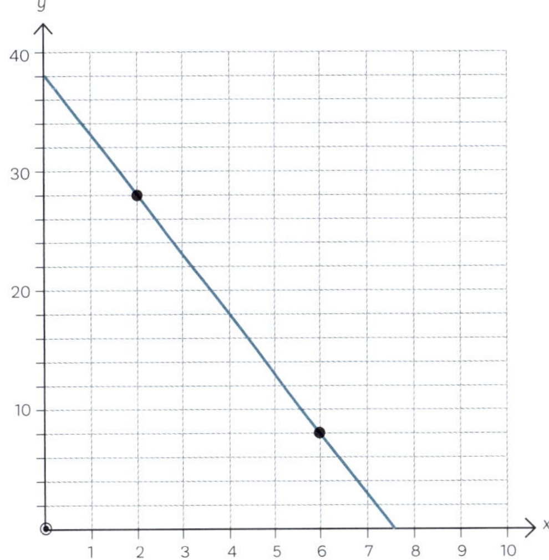

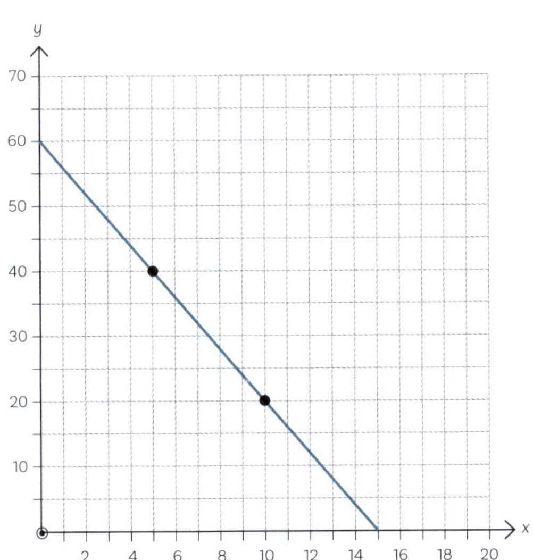

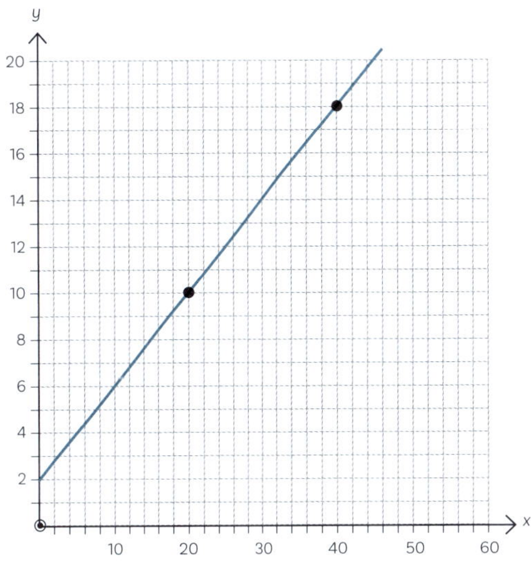

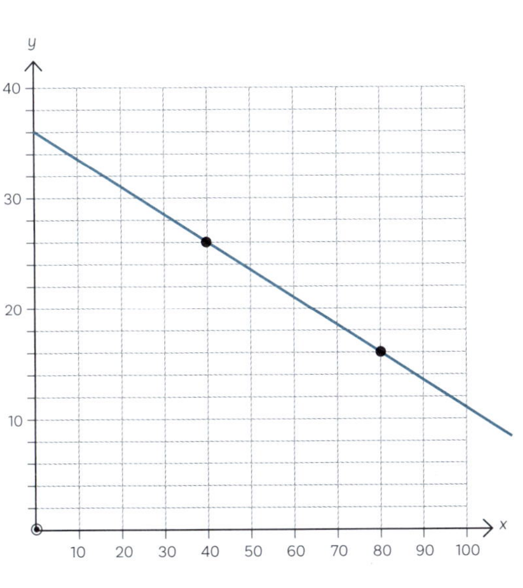

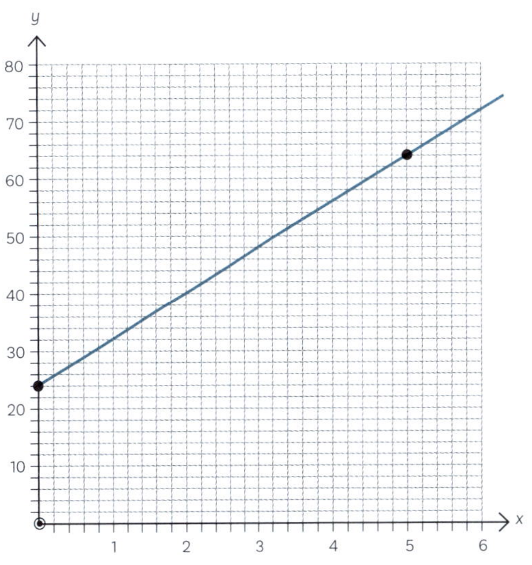

Drawing straight lines — continuous data

- **Continuous** data is unrounded **measured** data.
- It is shown as a **line** on a graph.

1 Plotting points using the equation

There are several ways of plotting graphs, but this method will work with *any* type of graph — lines and curves.

Example: Plot the line given by the equation $y = 2x - 3$.

Step 1: Create a table for values of x and y.

x	2x - 3	y	Point
0	2(0) - 3	-3	(0, -3)
1	2(1) - 3	-1	(1, -1)
2	2(2) - 3	1	(2, 1)
3	2(3) - 3	3	(3, 3)
4	2(4) - 3	5	(4, 5)

Calculate **at least three** points.

Step 2: Plot the points on a graph.

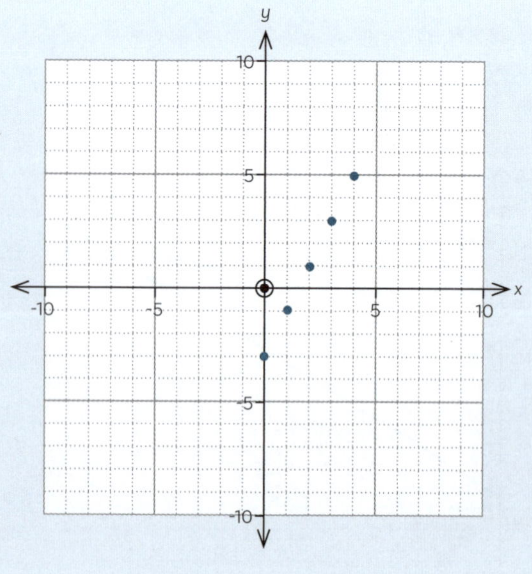

Step 3: Join the points with a **ruled** line.

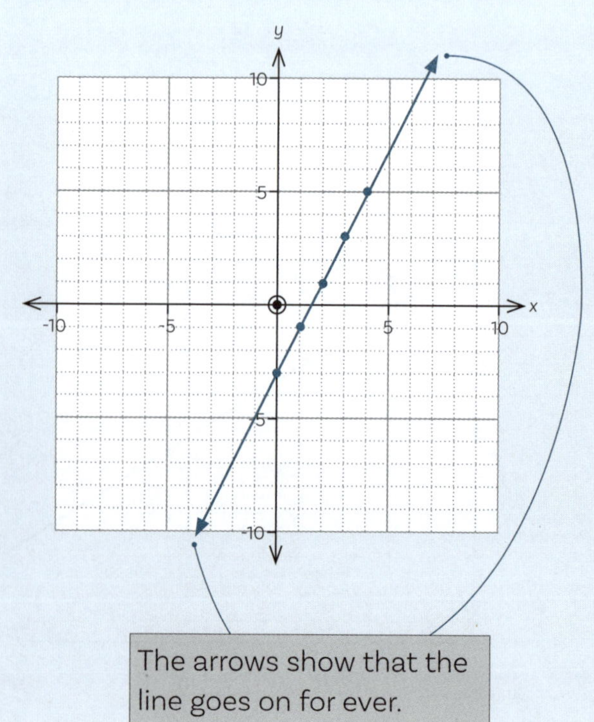

The arrows show that the line goes on for ever.

ISBN: 9780170484084

Complete the tables and draw the graph for each of the following.

1 $y = x + 3$

x	x + 3	y	Point
0	(0) + 3	3	(0, 3)
1			
2			
3			
4			

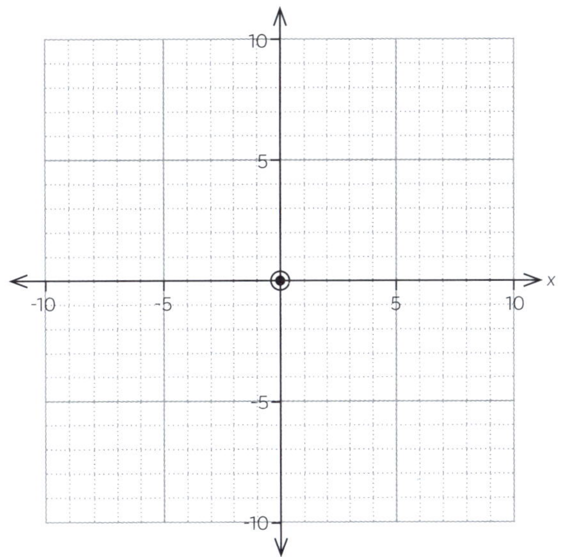

2 $y = 2x + 5$

x	2x + 5	y	Point
0			
1			
2			
3			
4			

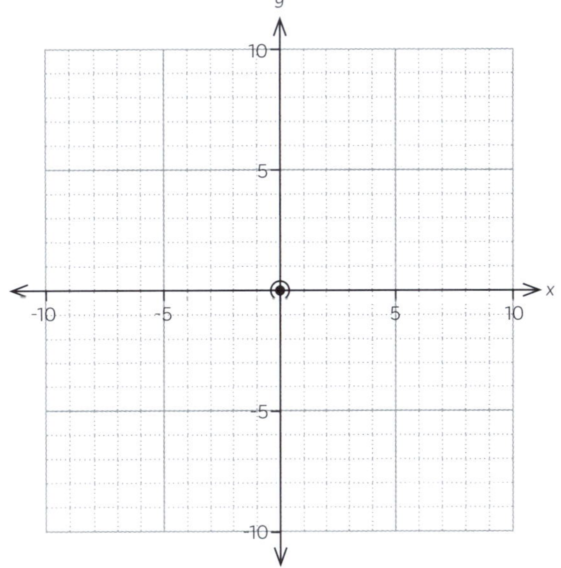

3 $y = 3x - 1$

x	3x - 1	y	Point
0			
1			
2			
3			
4			

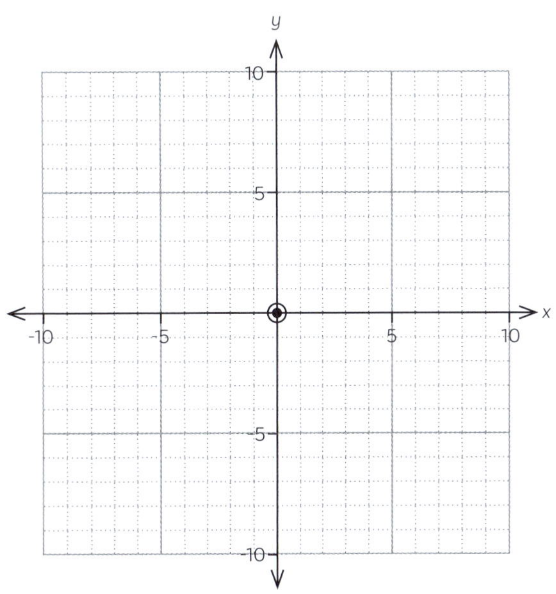

4 $y = -3x + 4$

x	-3x + 4	y	Point
0			
1			
2			
3			
4			

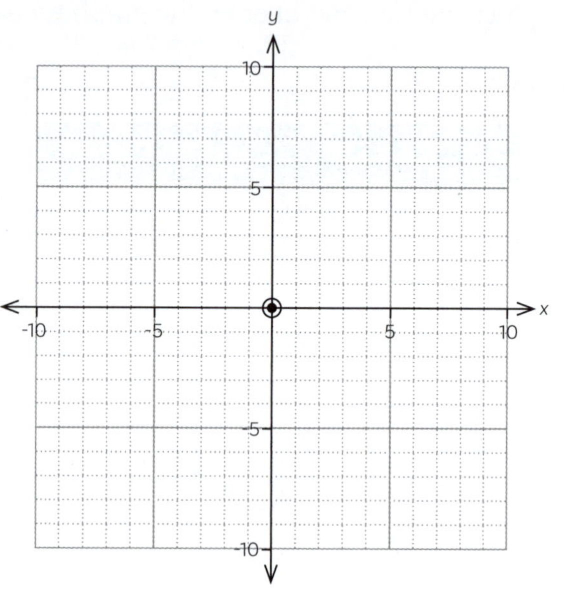

5 $y = 3$

x	3	y	Point
0			
1			
2			
3			
4			

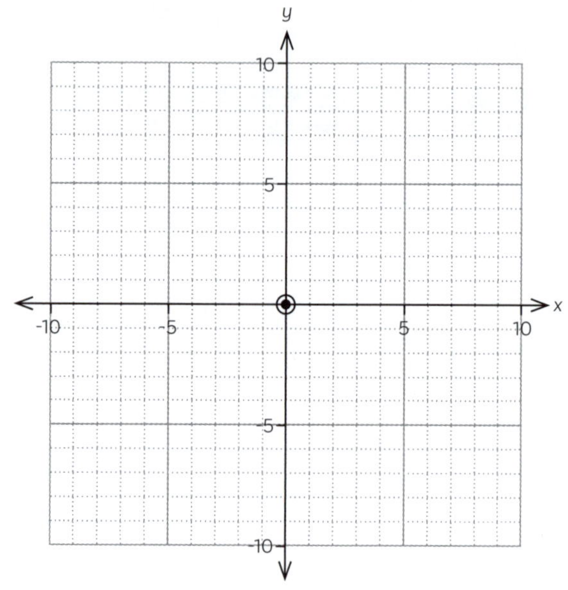

6 $y = -2x$

x	-2x	y	Point
0			
1			
2			
3			
4			

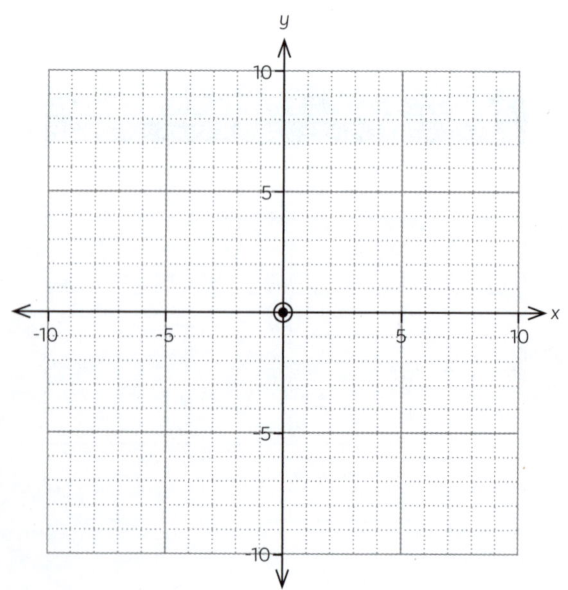

ISBN: 9780170484084

2 Using the y-intercept and the gradient

If you are given the equation of a graph, you can convert it into the form $y = mx + c$ before drawing the graph.

m = gradient = $\frac{\text{rise}}{\text{run}}$ — $y = mx + c$ — c = y-intercept

Examples:

1 Draw the graph of $y = -\frac{1}{2}x - 3$.

Step 1: Plot the y-intercept; in this case the y-intercept = **−3**.

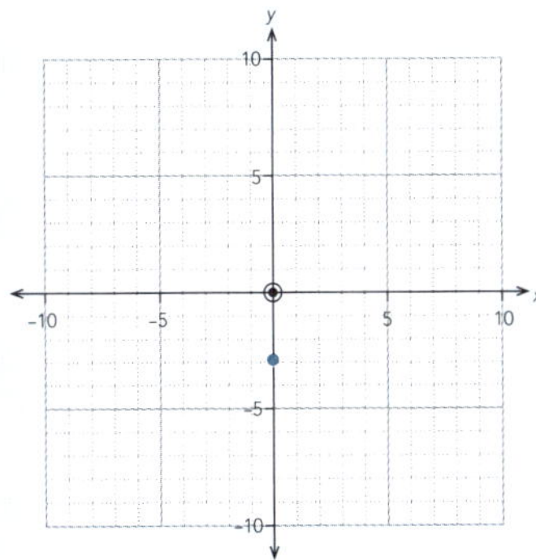

Step 2: From the intercept, plot at least three points along the gradient, in this case $\frac{-1}{2}$.

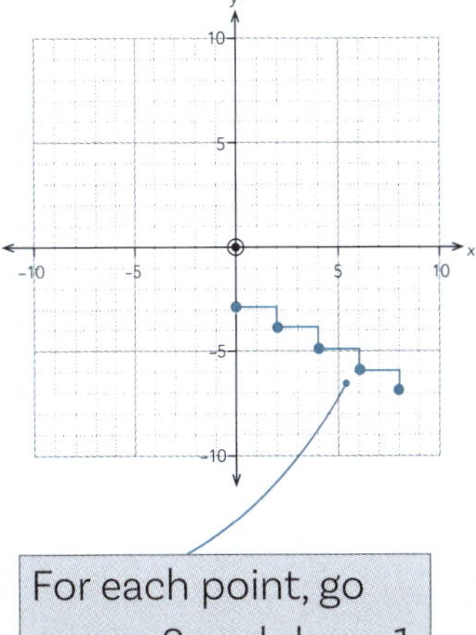

Step 3: Join the points.

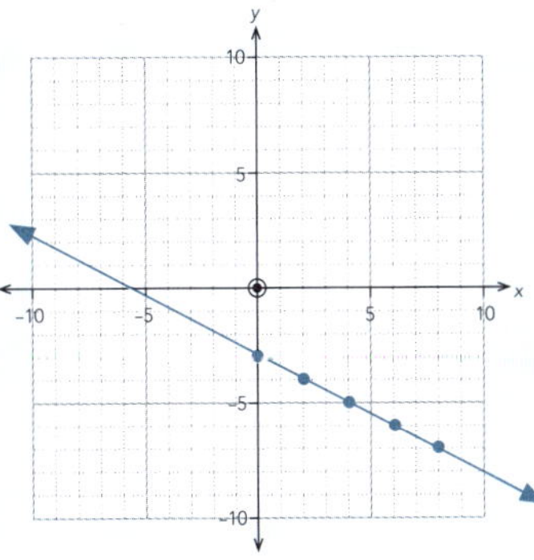

2 Draw the graph of $y = 60x + 200$.

Step 1: Plot the y-intercept; in this case the y-intercept = **200**.

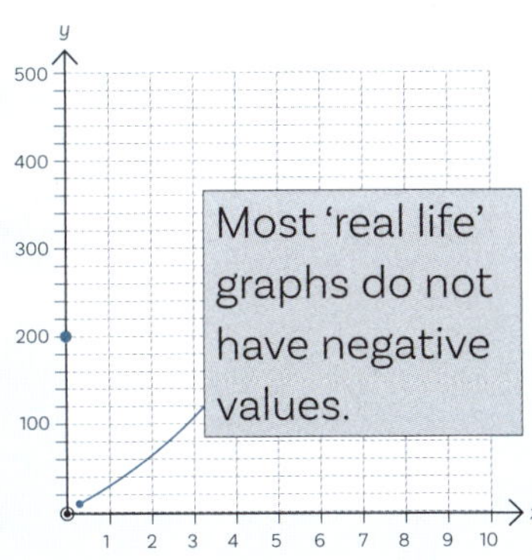

Step 2: From the intercept, plot at least three points along the gradient, in this case $\frac{60}{1}$.

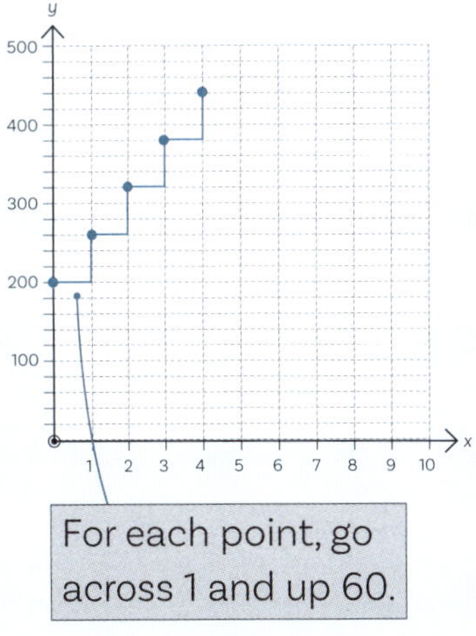

Step 3: Join the points.

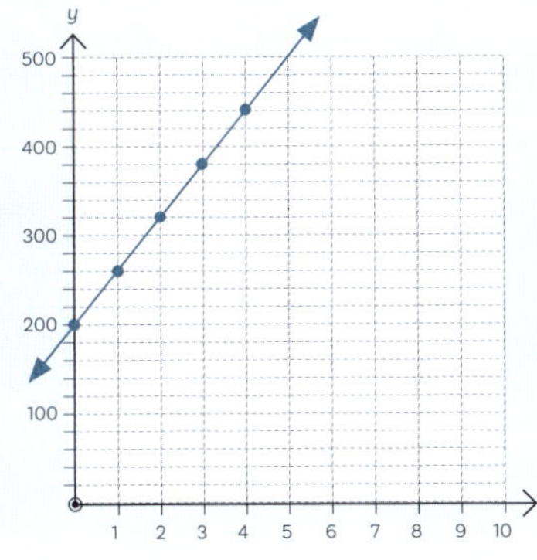

Note: Depending on the context, the line mightt stop at the axis.

ISBN: 9780170484084

Draw the following lines.

1 $y = x + 4$

2 $y = 2x - 5$

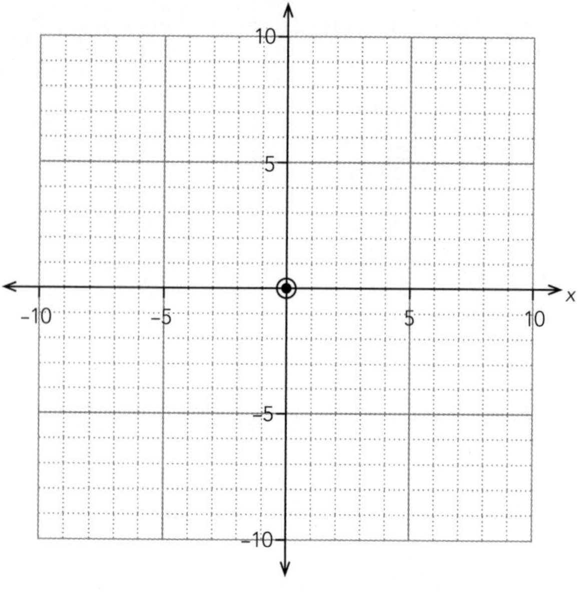

3 $y = -3x + 2$

4 $y = \frac{1}{3}x - 4$

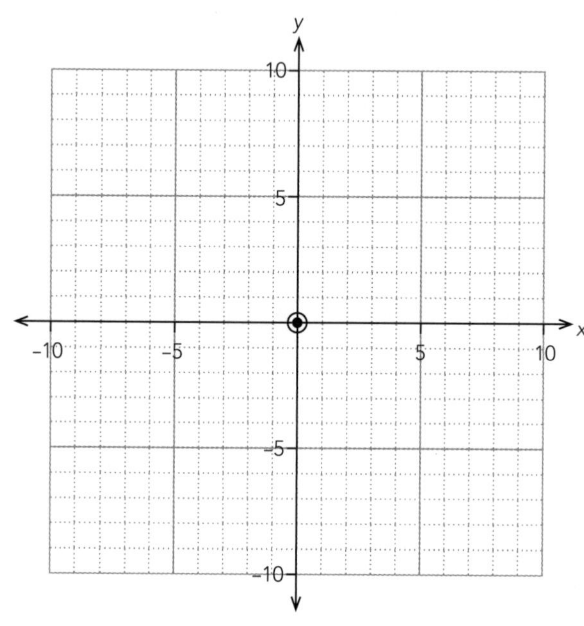

5 $y = -\frac{1}{4}x - 5$

6 $y = -3x$

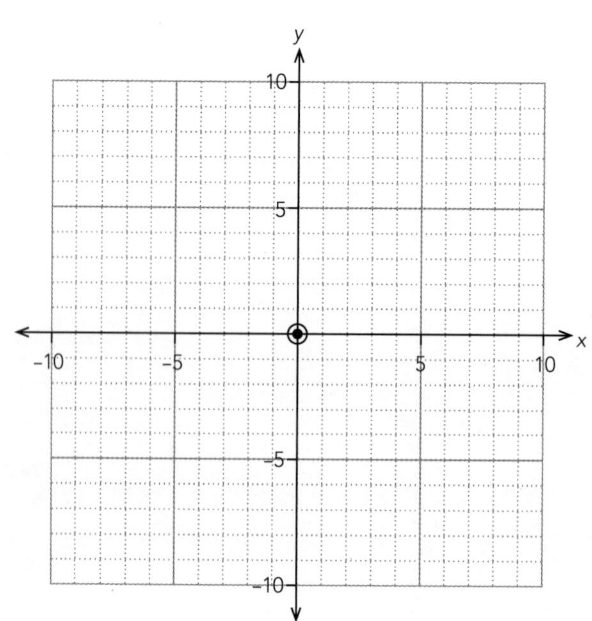

ISBN: 9780170484084

7 $y = 100x + 200$

8 $y = -50x + 500$

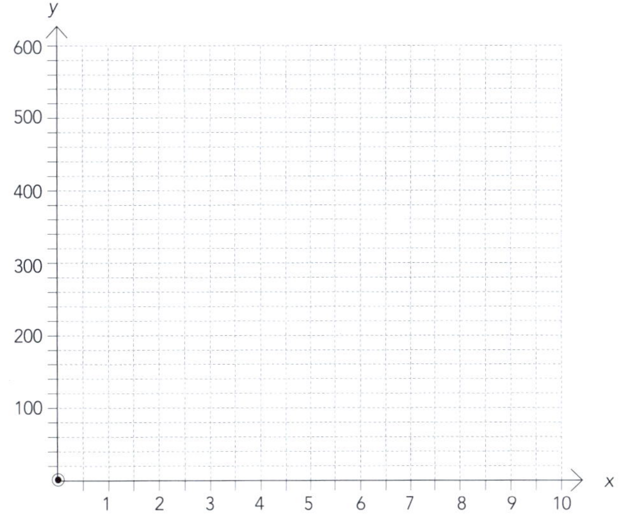

9 $y = 6x + 12$

10 $y = -4x + 36$

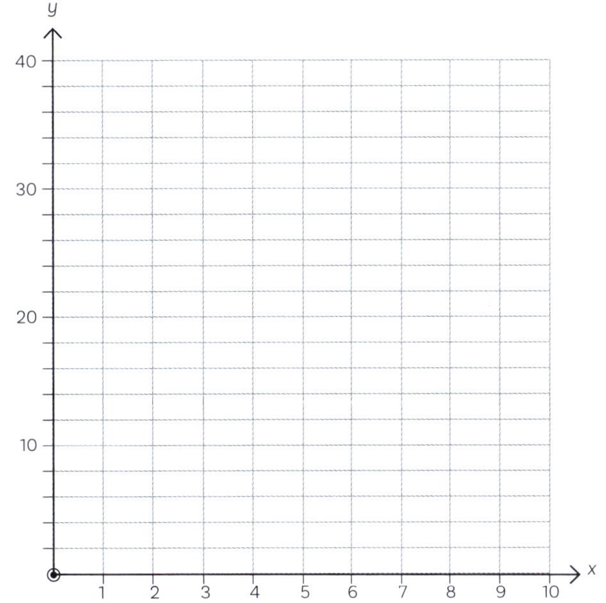

11 $y = -0.5x + 18$

12 $y = 0.2x + 5$

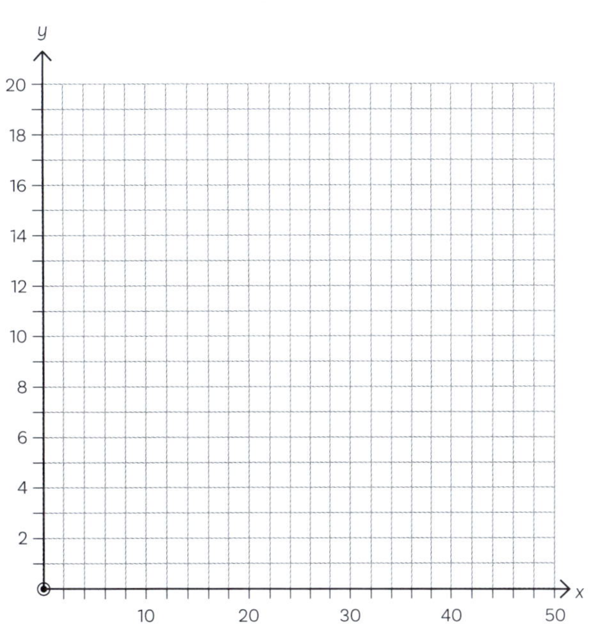

3 Using the x- and y-intercepts

- Often you will need to draw the graph of an equation that is not in $y = mx + c$ form.
- If it is in the form $ax + by = c$, it is usually easier to use the **intercept-intercept** method in order to draw the graph.

Examples:

1 Draw the graph of $6x - 3y = 12$.

Step 1: Make $x = 0$.
Then $6(0) - 3y = 12$
$\therefore\ y = \mathbf{-4}$
Plot the point (0, **-4**).

Step 2: Make $y = 0$.
Then $6x - 3(0) = 12$
$\therefore\ x = \mathbf{2}$
Plot the point (**2**, 0).

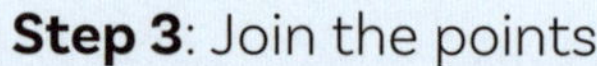

Step 3: Join the points.

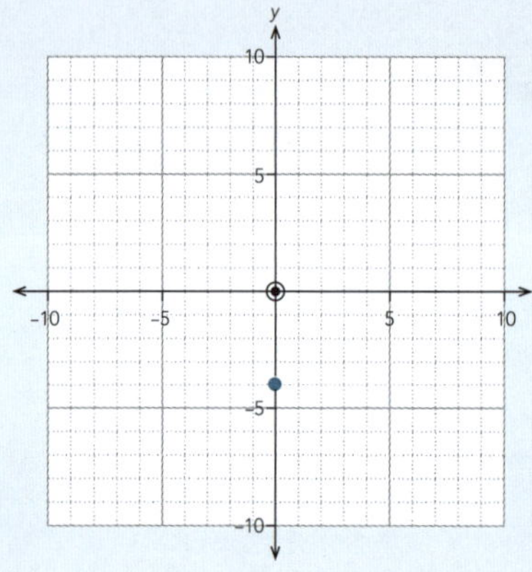

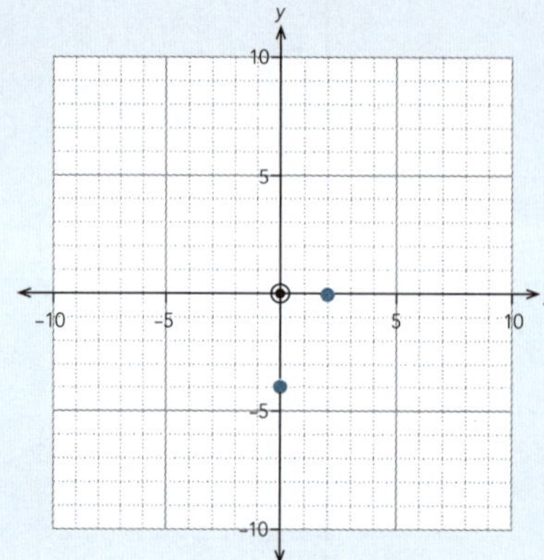

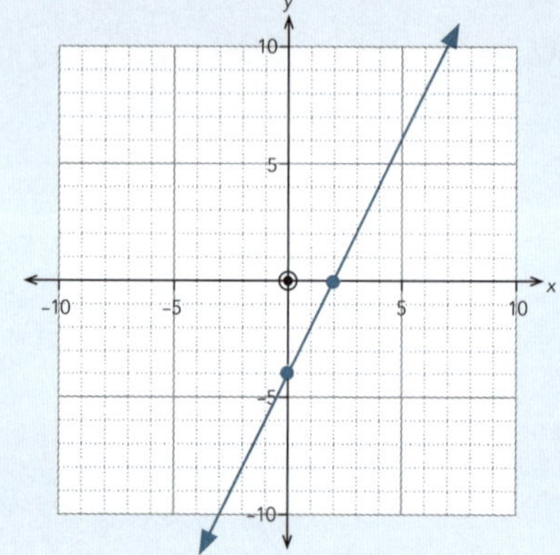

2 Draw the graph of $3x + 2y = 90$.

Step 1: Make $x = 0$.
Then $3(0) + 2y = 90$
$\therefore\ y = \mathbf{45}$
Plot the point (0, **45**).

Step 2: Make $y = 0$.
Then $3x + 2(0) = 90$
$\therefore\ x = \mathbf{30}$
Plot the point (**30**, 0).

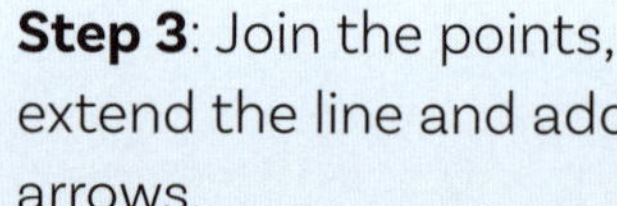

Step 3: Join the points, extend the line and add arrows.

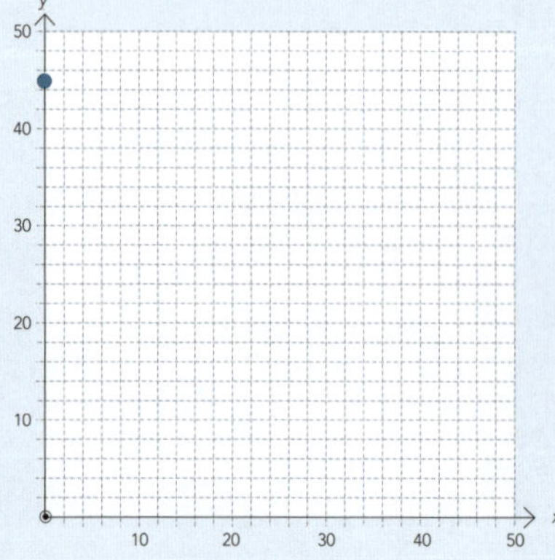

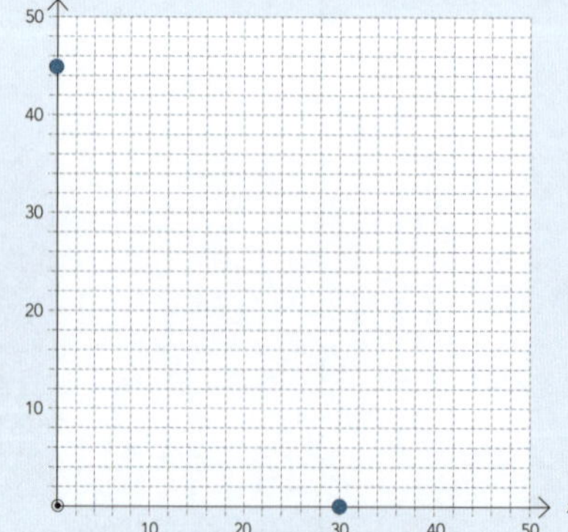

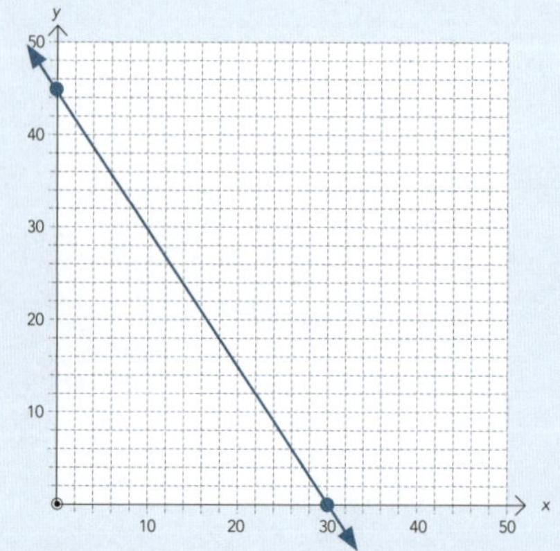

ISBN: 9780170484084

Draw the following lines.

1 $4x - 2y = 20$

$x = 0 \Rightarrow y =$ ____ $y = 0 \Rightarrow x =$ ____

2 $3x + 4y = 12$

$x = 0 \Rightarrow y =$ ____ $y = 0 \Rightarrow x =$ ____

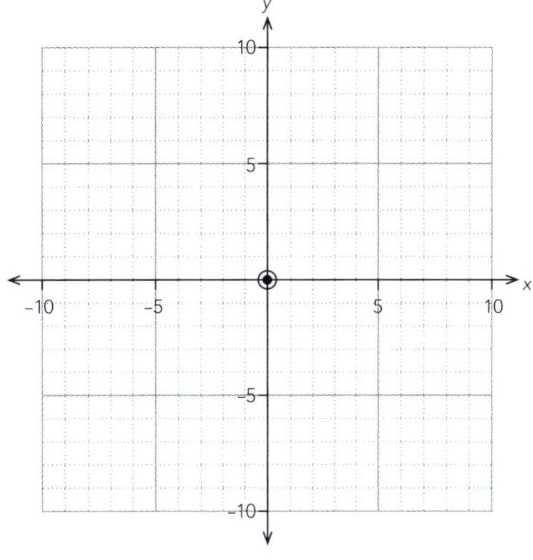

3 $3x + 2y = 18$

4 $3x - 6y = 24$

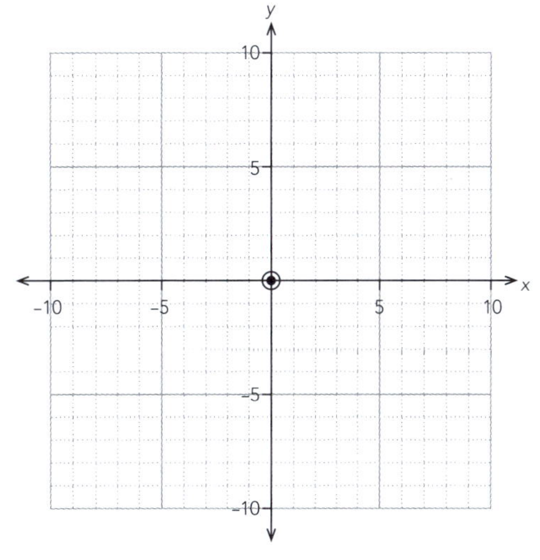

5 $3y + 5x = 150$

6 $4x + 5y = 200$

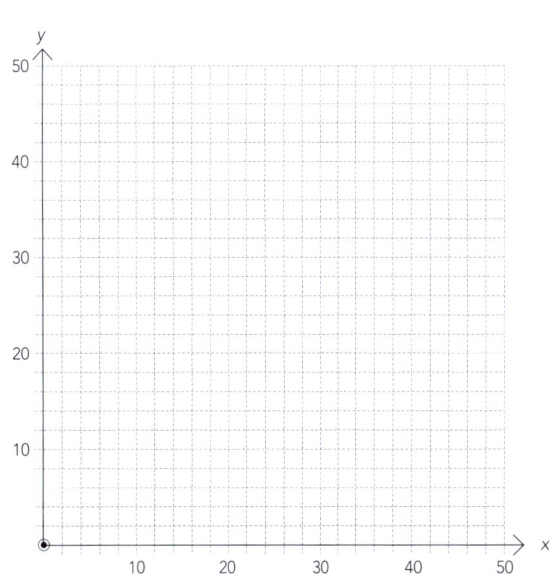

4 Horizontal and vertical lines

Be very careful when drawing these – it's easy to get them the wrong way around.

Examples:

1 Draw the line $x = -3$.

Step 1: Plot at least three points where $x = -3$, e.g. (-3, 0), (-3, 5), (-3, -4).

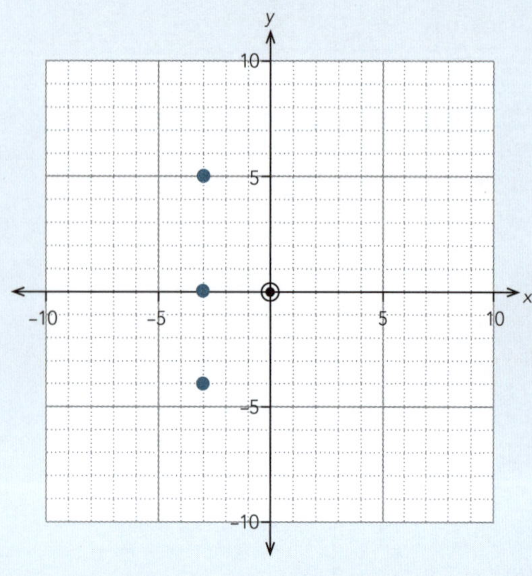

Step 2: Join the points.

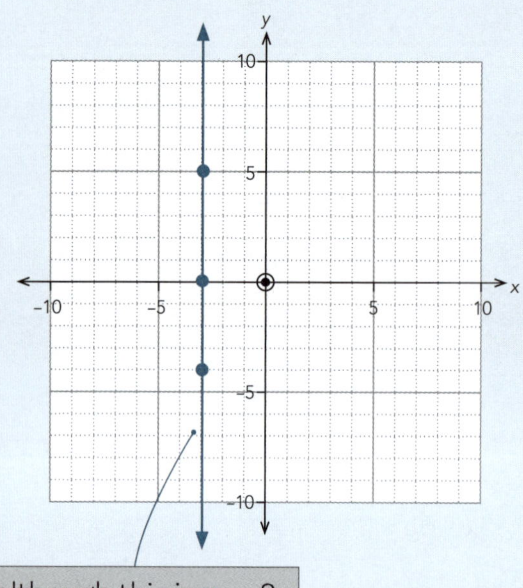

Notice that although this is **x** = -3, it is parallel with the **y**-axis.

2 Draw the line $y = 4$.

Step 1: Plot at least three points where $y = 4$, e.g. (0, 4), (2, 4), (-3, 4).

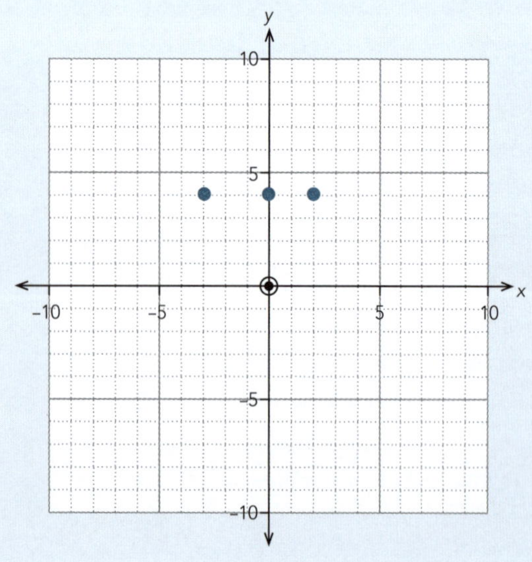

Step 2: Join the points.

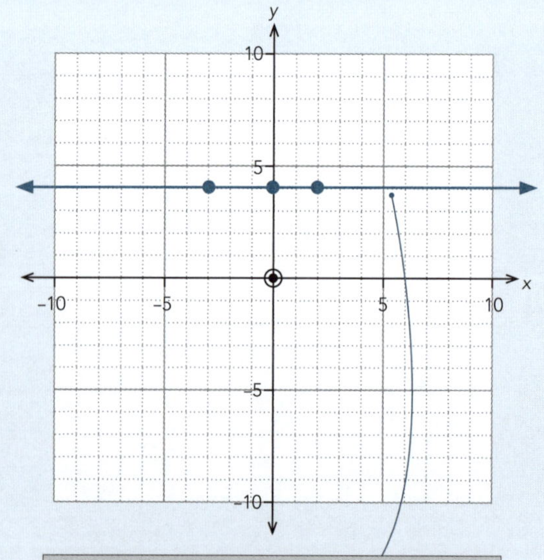

Again, although this is **y** = 4, it is parallel with the **x**-axis.

ISBN: 9780170484084

Draw the following lines.

1 $x = 3$ and $y = 5$

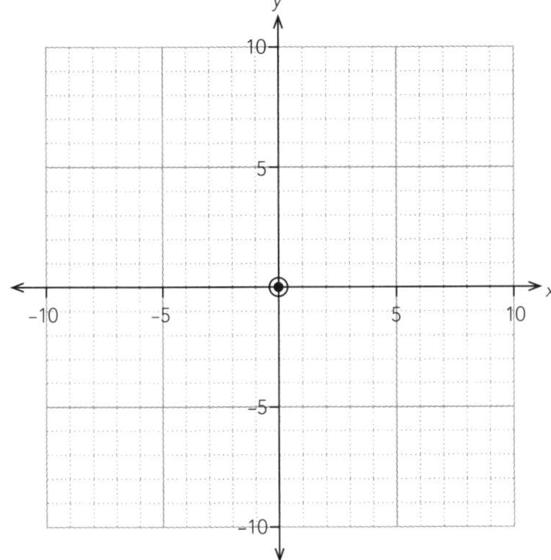

2 $y = -2$ and $x = 7$

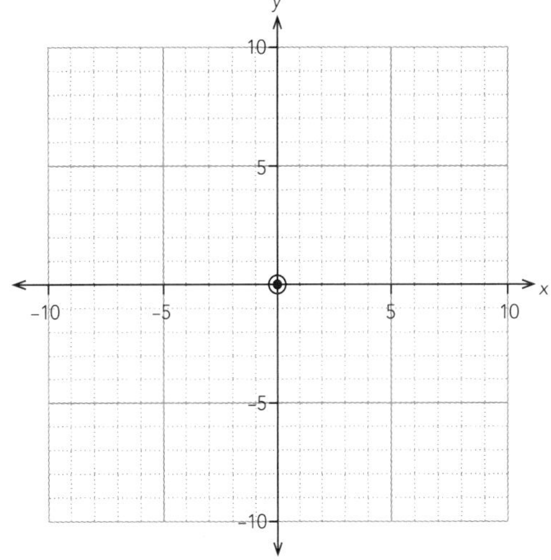

3 $x = -1$ and $y = 0$

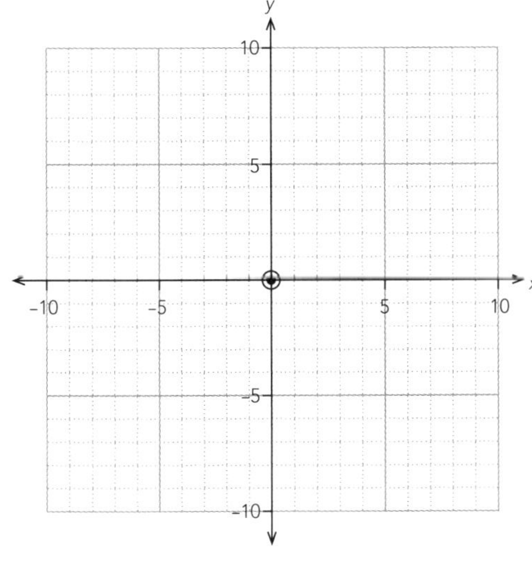

4 $y = 6$ and $x = 0$

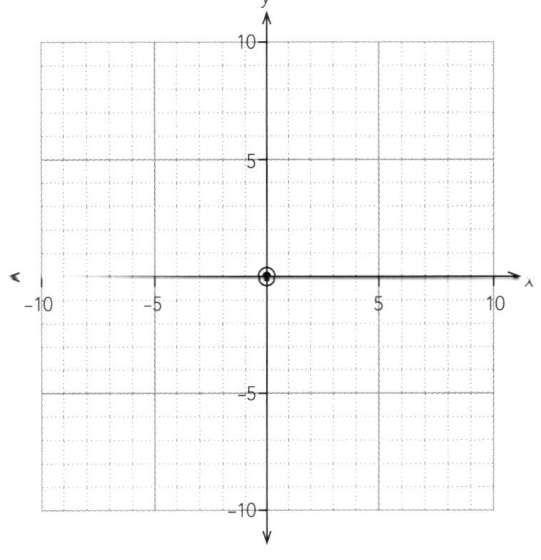

5 $x = 24$ and $y = 8$

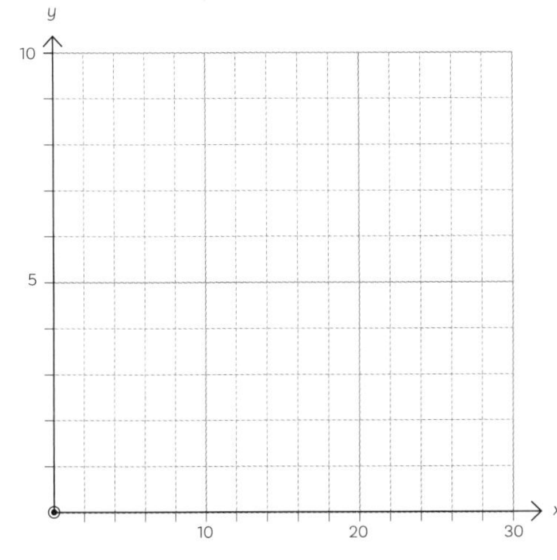

6 $x = 14$ and $y = 18$

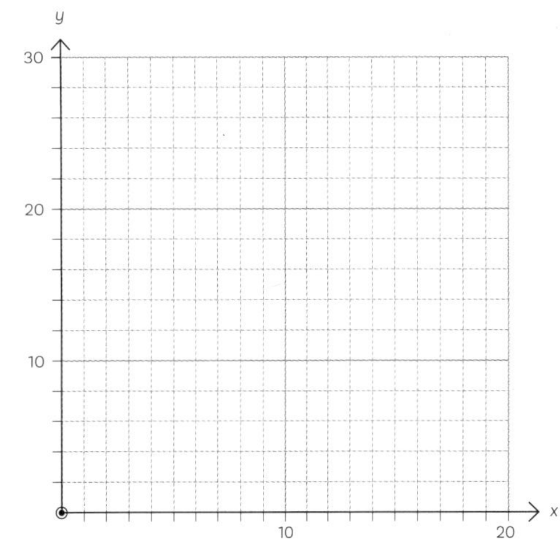

Writing equations from graphs

1 Using the gradient and the y-intercept

Use $\mathbf{y = mx + c}$ as a template for your equation.

Example 1: Write the equation for this line.

Step 1: Find the y-intercept (c).

$c = -4$

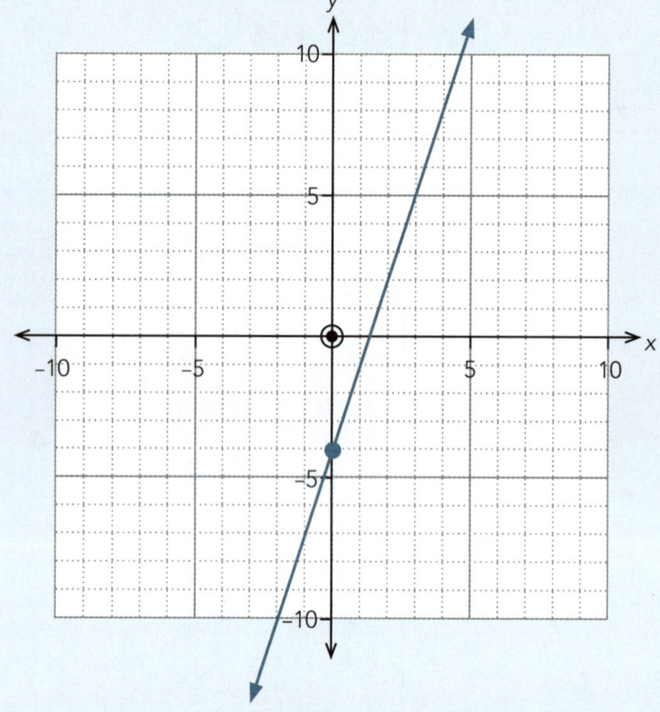

Step 2: Work out the gradient (m).

Draw a right-angled triangle between any two points through which the line passes.

$$m = \frac{\text{rise}}{\text{run}} = \frac{+12}{+4} = 3$$

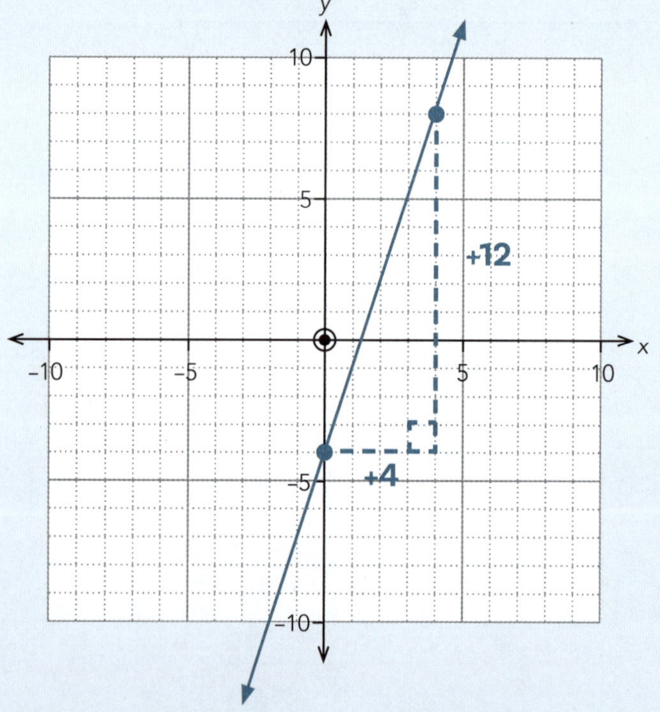

Step 3: Substitute into $y = mx + c$.

$\mathbf{y = mx + c}$

m = 3

c = −4

So: $\mathbf{y = 3x - 4}$

Step 4: Check. Substitute at least one of the points on the graph into the equation.

At (4, 8): $y = 3 \times 4 - 4 = 8$ ✓

ISBN: 9780170484084

Example 2: Write the equation for this line.

Step 1: Find the y-intercept (c).

c = 90

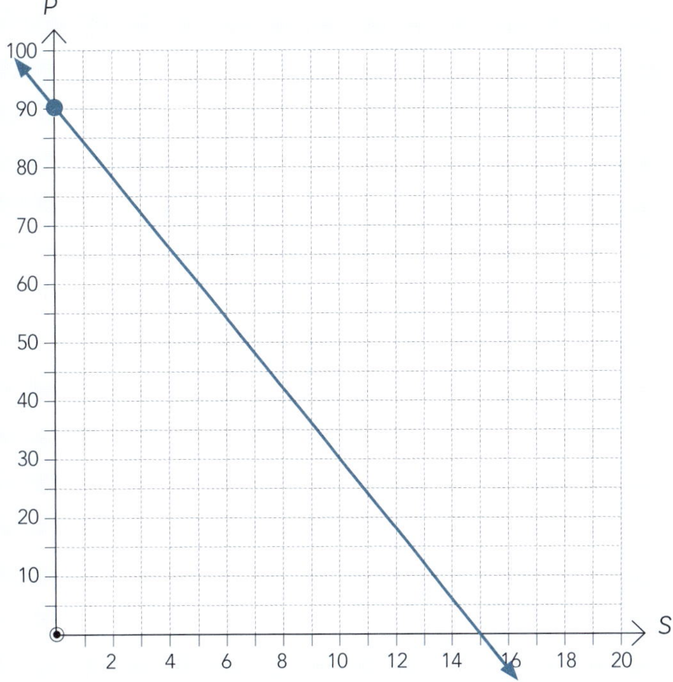

Step 2: Work out the gradient (m).

Draw a right-angled triangle between any two points through which the line passes.

$$m = \frac{\text{rise}}{\text{run}} = \frac{-60}{+10} = -6$$

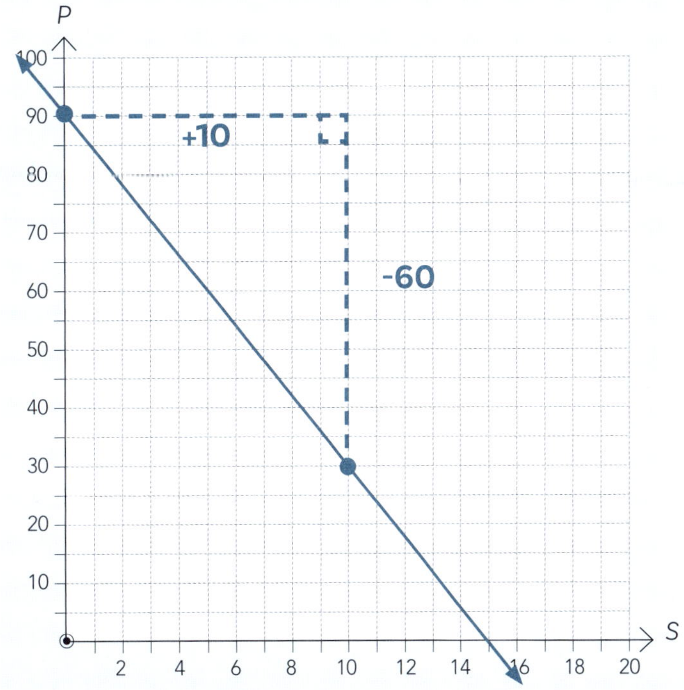

Step 3: Substitute into $y = mx + c$.

$y = mx + c$

m = -6   c = 90

So: $y = -6x + 90$

But, in this case, we have P on the y-axis and S on the x-axis. So the equation has to be:

$$P = -6S + 90$$

Step 4: Check. Substitute at least one of the points on the graph into the equation.

At (10, 30): P = −6 x 10 + 90 = 30 ✓

ISBN: 9780170484084

Write equations for the following lines.

1

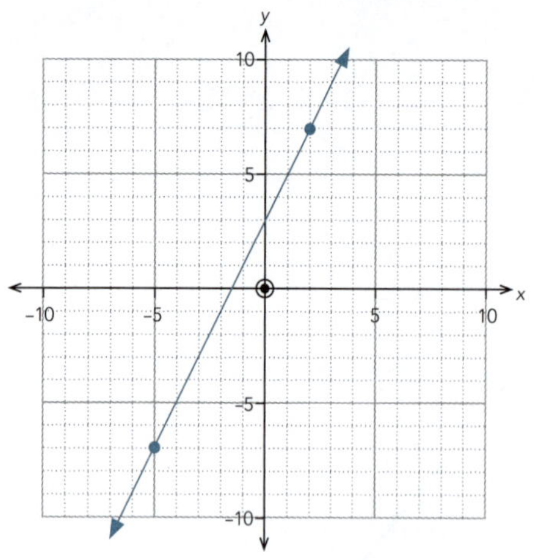

y-intercept = ______ Gradient = ______

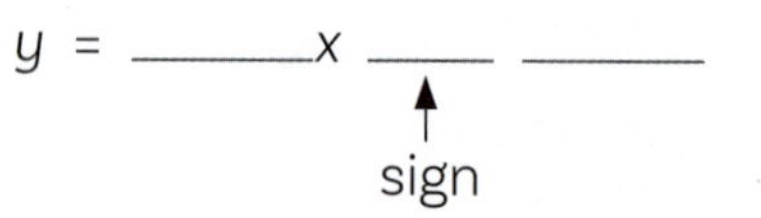

2

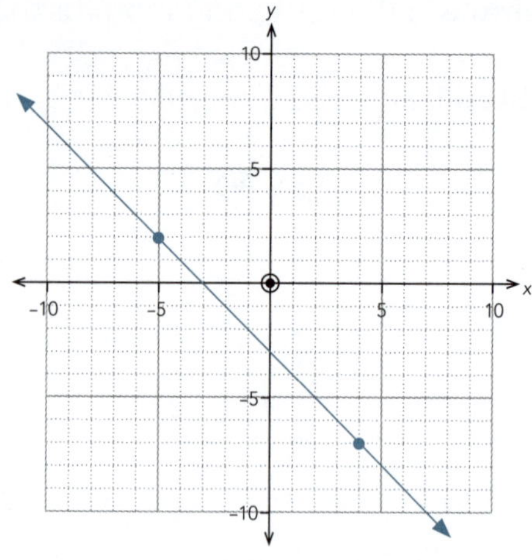

y-intercept = ______ Gradient = ______

y = ______x ______ ______

sign

3

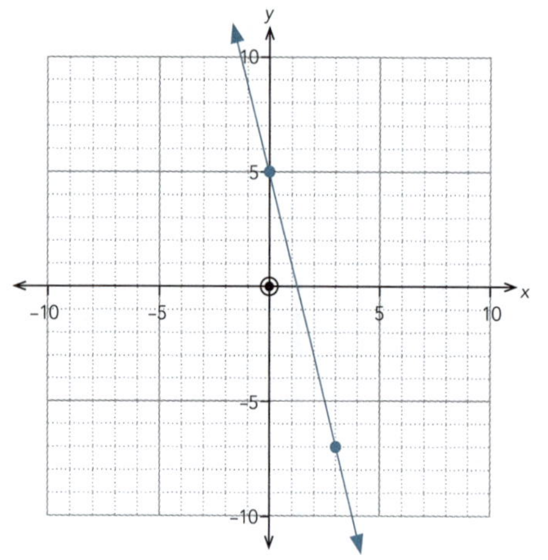

y-intercept = ______ Gradient = ______

y = ____________

4

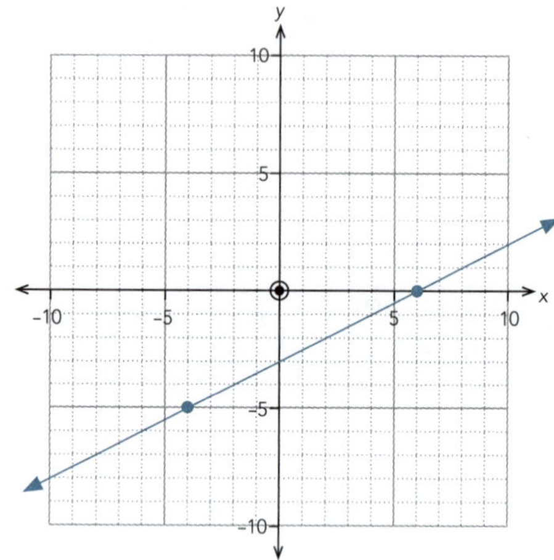

y-intercept = ______ Gradient = ______

y = ____________

5

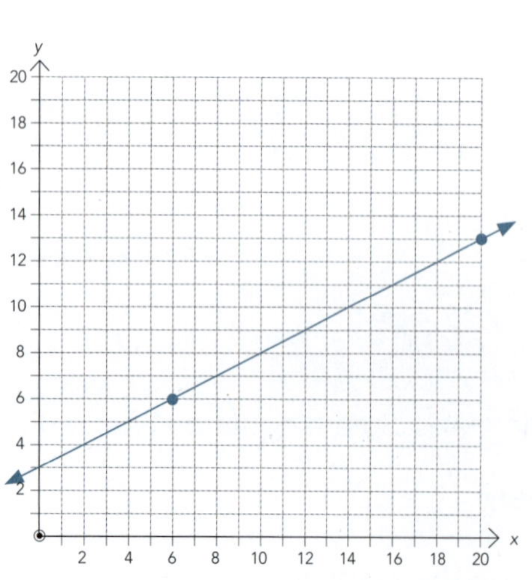

y = ____________

6

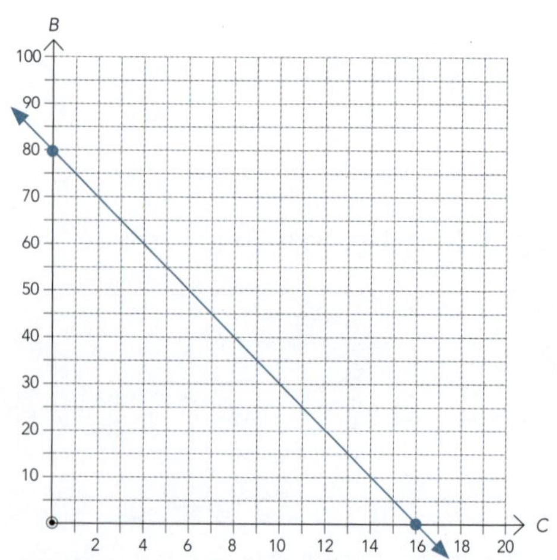

B = ____________

 ISBN: 9780170484084

7

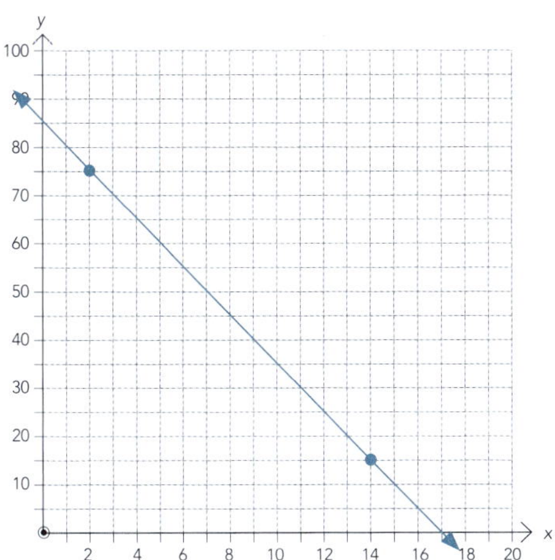

$y =$ ______________________

8

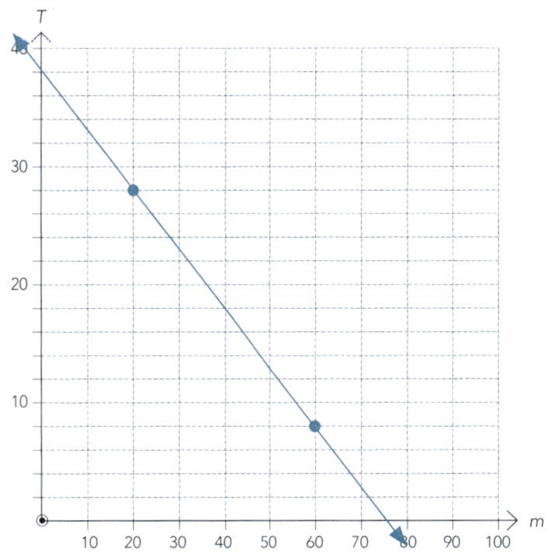

$T =$ ______________________

9

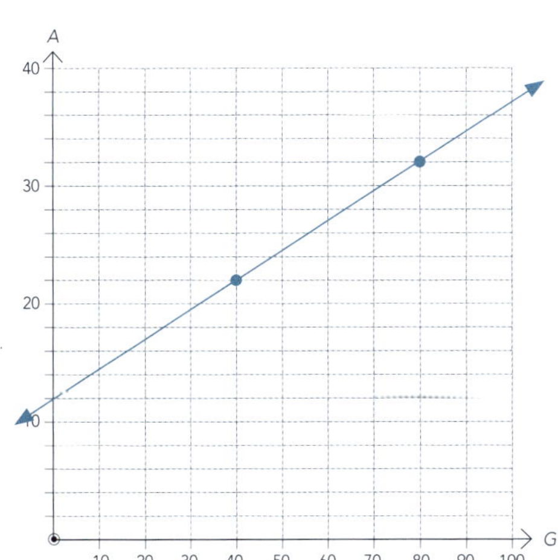

$A =$ ______________________

10

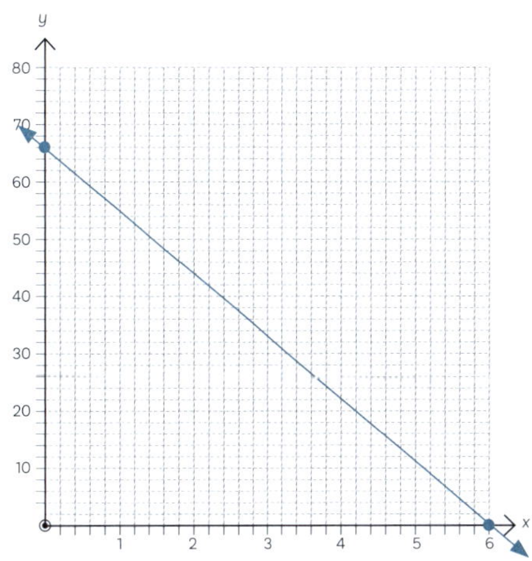

$y =$ ______________________

11

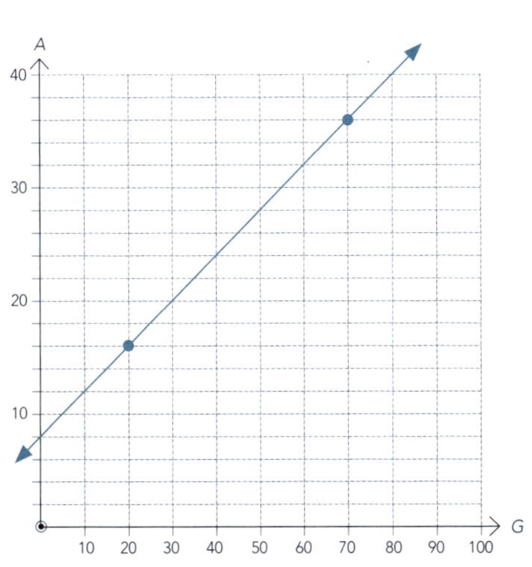

$A =$ ______________________

12

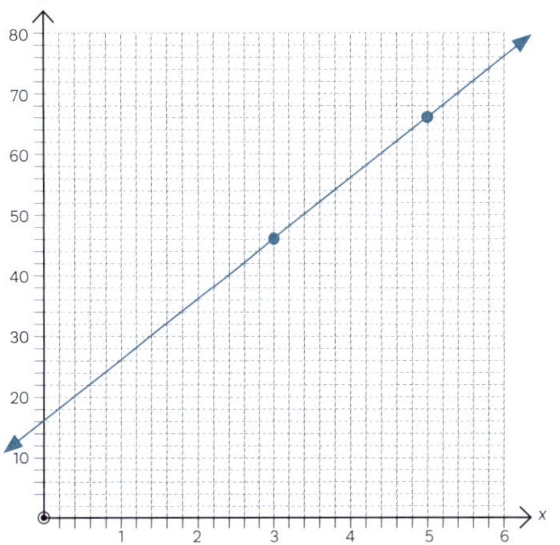

$y =$ ______________________

2 Using two points

Label the two points (x_1, y_1) and (x_2, y_2), then:

1 Substitute the coordinates into $\mathbf{m = \frac{rise}{run} = \frac{y_2 - y_1}{x_2 - x_1}}$ in order to find the gradient.

2 Substitute the gradient and the coordinates into $\mathbf{y - y_1 = m(x - x_1)}$ in order to find the equation.

Example 1: Find the equation of the line that passes through (5, 1) and (6, 4).

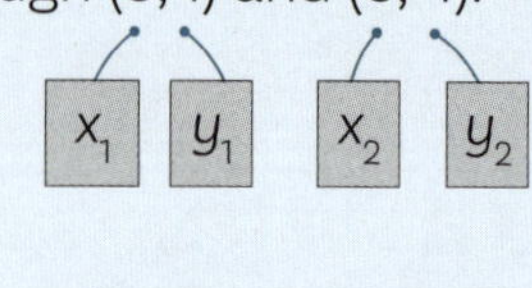

Step 1: Find the gradient.

$$\mathbf{m} = \frac{y_2 - y_1}{x_2 - x_1} = \frac{4 - 1}{6 - 5} = \frac{3}{1} = 3$$

Step 2: Substitute this, along with x_1 and y_1, into $\mathbf{y - y_1 = m(x - x_1)}$.

$$\therefore y - 1 = 3(x - 5)$$

$$y - 1 = 3x - 15 \quad \mathbf{(+1)}$$

$$y = 3x - 14$$

Step 3: Check. Substitute at least one of the points into the equation.

At (5, 1): $y = 3(5) - 14 = 1$ ✓

Example 2: Find the equation of the line that passes through (4, -3) and (-6, 2).

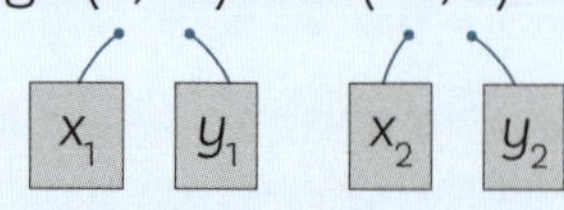

Step 1: Find the gradient.

$$\mathbf{m} = \frac{y_2 - y_1}{x_2 - x_1} = \frac{2 - (-3)}{-6 - 4} = -\frac{1}{2}$$

Step 2: Substitute this, along with x_1 and y_1, into $\mathbf{y - y_1 = m(x - x_1)}$.

$$\therefore y - (-3) = -\frac{1}{2}(x - 4)$$

$$y + 3 = -\frac{1}{2}x + 2 \quad \mathbf{(-3)}$$

$$y = -\frac{1}{2}x - 1$$

Step 3: Check. Substitute at least one of the points into the equation.

At (4, -3): $y = -\frac{1}{2} \times 4 - 1 = -3$ ✓

 ISBN: 9780170484084

Calculate the equation of the line that connects the following pairs of points.

1 (1, 5) and (3, 9)

2 (2, 8) and (4, 9)

3 (5, 1) and (9, 5)

4 (-2, 1) and (1, 7)

5 (0, 3) and (1, 1)

6 (8, -2) and (1, 5)

7 (-4, 3) and (6, -2)

8 (1, -7) and (-2, 5)

GEOMETRY

Naming angles

- Angles can be named in two ways.

1 Using the **three** usually upper case letters **outside** each vertex.

This angle is ∠DEF or ∠FED.

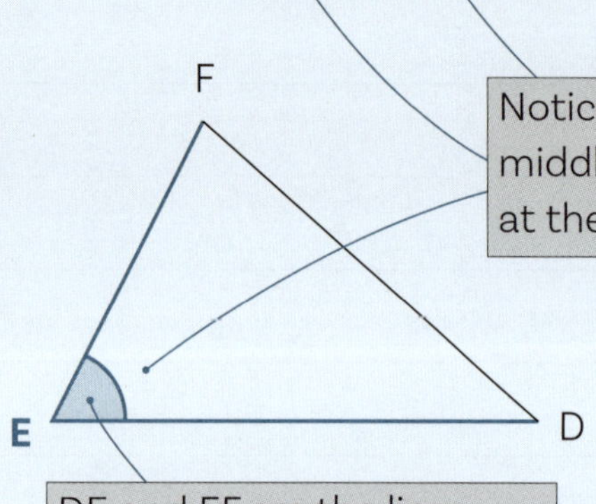

2 Using the **one** usually lower case letter **inside** the angle.

This angle is ∠y.

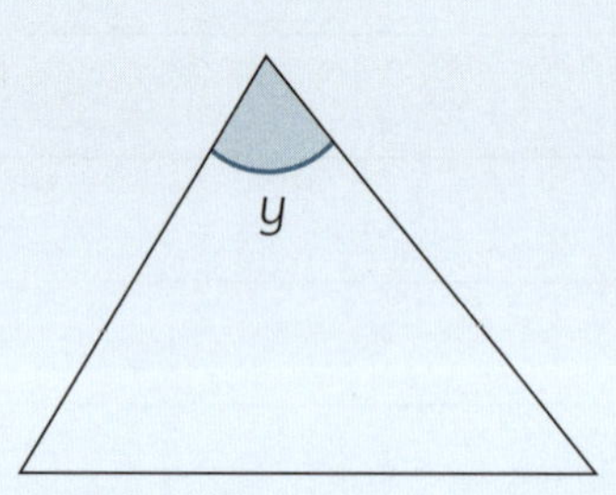

Match the angle names to the angles in the diagrams.

∠CBA	∠CAD	∠A	∠CAB	∠CBD
∠DAB	∠BCD	∠b	∠a	∠BDC

1

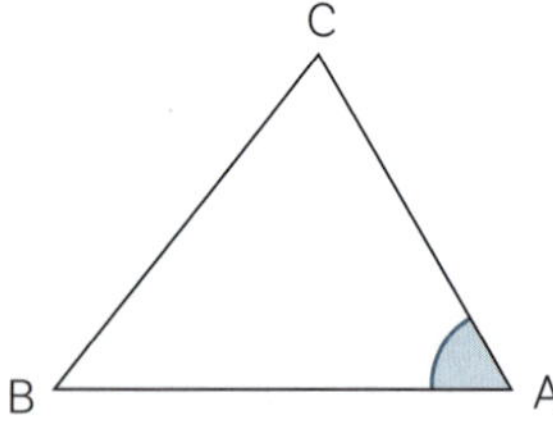

____

2

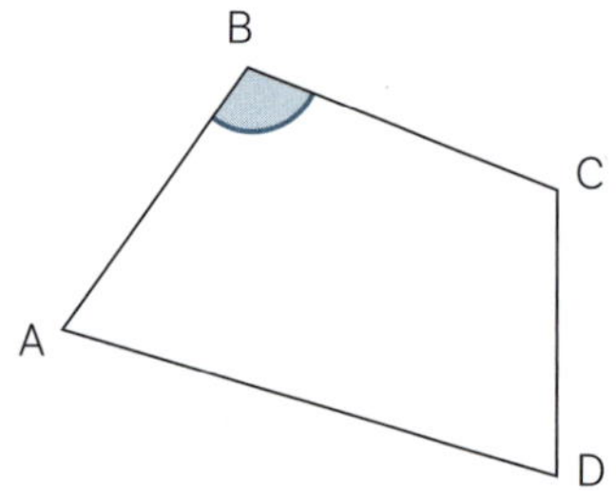

____

3

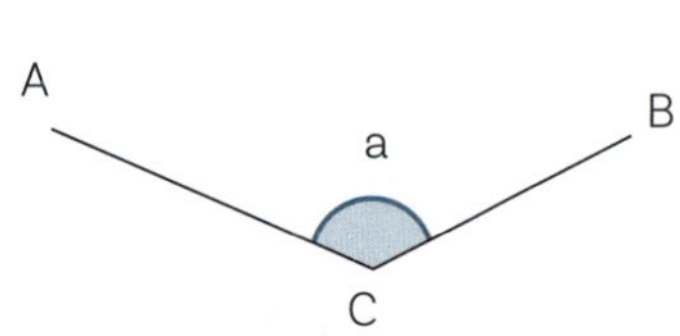

____

4

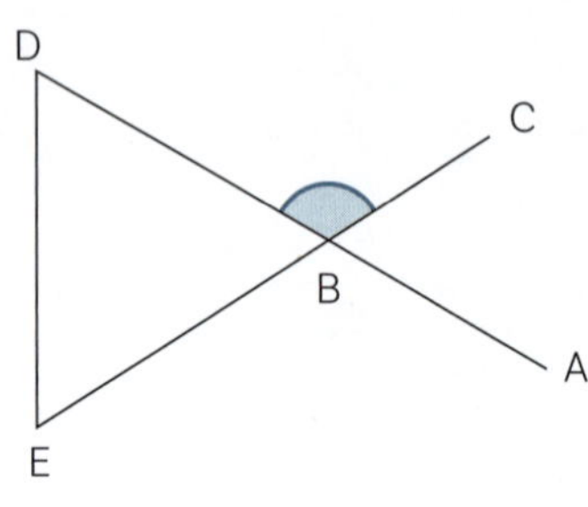

____

5

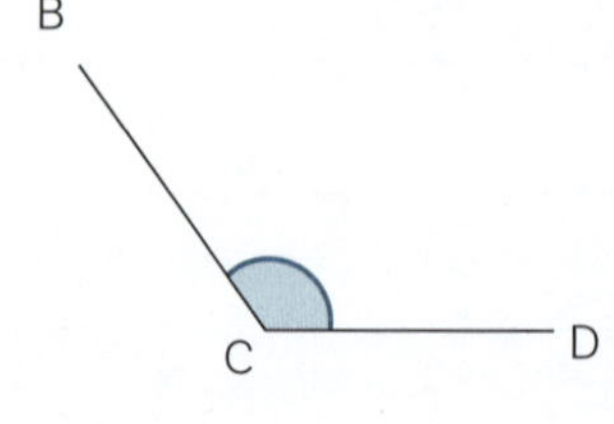

____

6

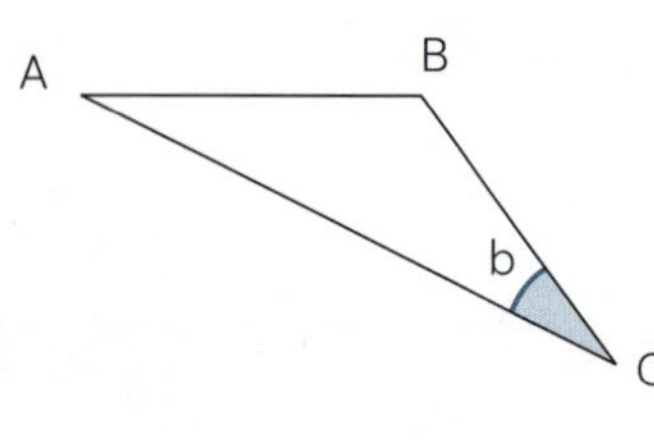

____

ISBN: 9780170484084

Types of angles

Type	Picture	Description
Acute angle		An angle between 0° and 90°.
Right angle		An angle that is 90°.
Obtuse angle		An angle between 90° and 180°.
Straight angle		An angle that is 180°.
Reflex angle		An angle between 180° and 360°.

Write down the type of each angle.

1

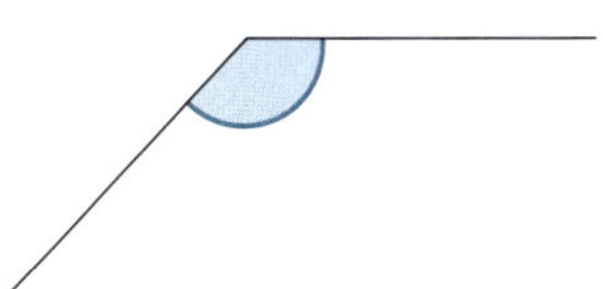

____

2

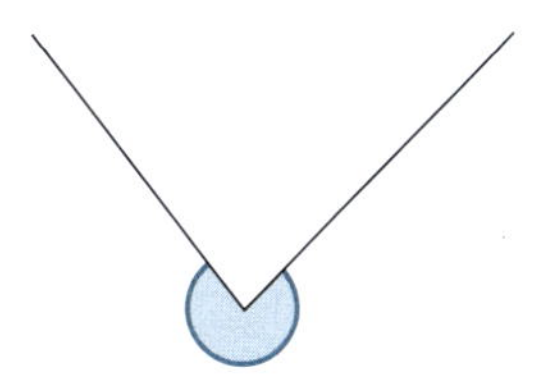

____

3

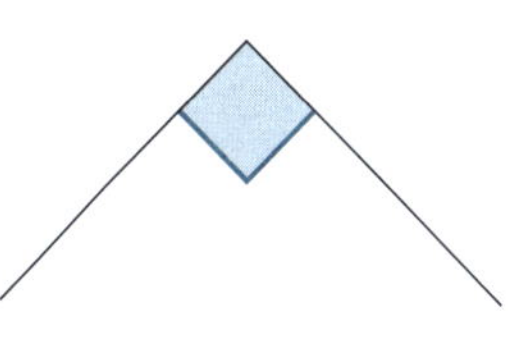

____

4

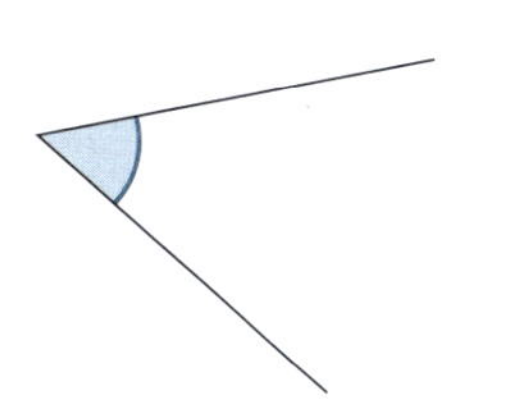

____

5

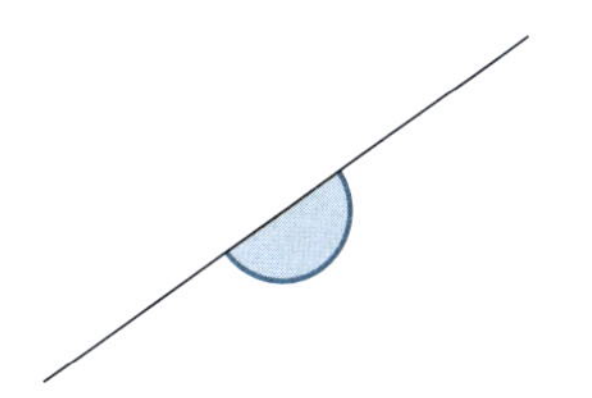

____

6

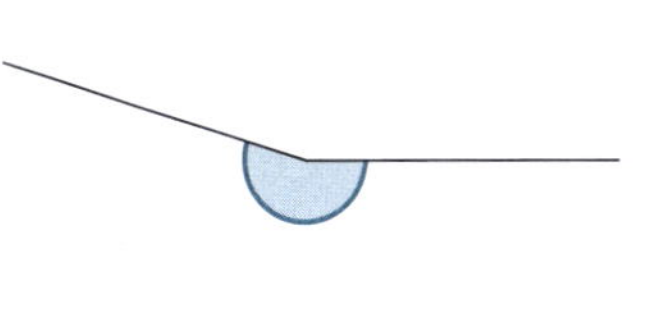

____

Angle rules

Relationship	Reason
$a + b = 180°$	Angles on a line add to 180°. (∠s on a line = 180°)
$a + b + c = 360°$	Angles at a point add to 360°. (∠s at a point = 360°)
$a = b$	Vertically opposite angles are equal. (vert opp ∠s =)
$a + b + c = 180°$	Angles in a triangle add to 180°. (∠s in Δ = 180°)
$a + b = c$	The exterior angle of a triangle = the sum of the interior opposite angles. (ext ∠ of Δ = sum of int opp ∠s)

The expressions in brackets are the short ways of writing the reasons.

 ISBN: 9780170484084

Calculate the missing angles and give the reason(s) used. If your reasons are different from those in the answers, check with your teacher.

1

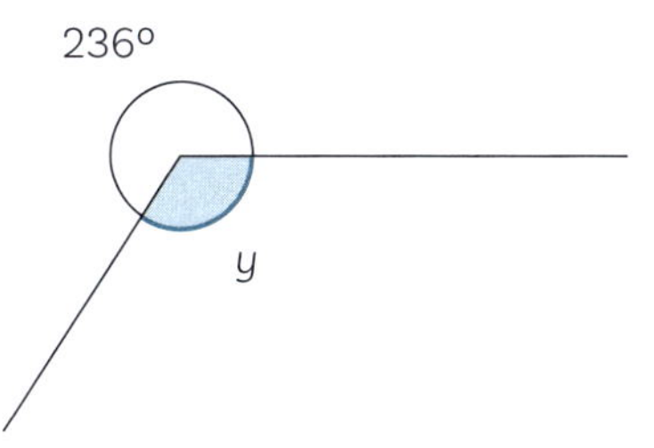

Reason(s): ______

2

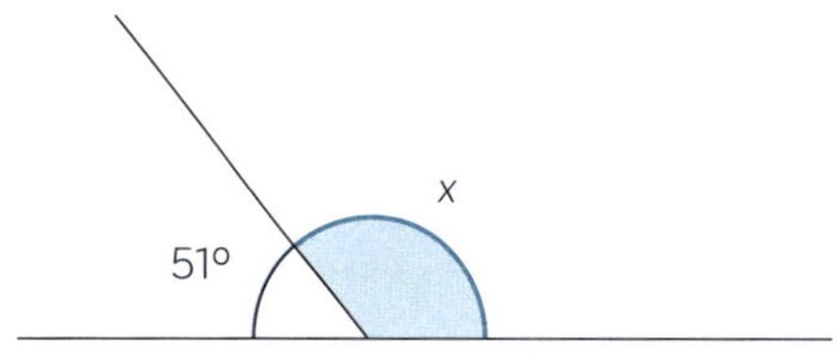

Reason(s): ______

3

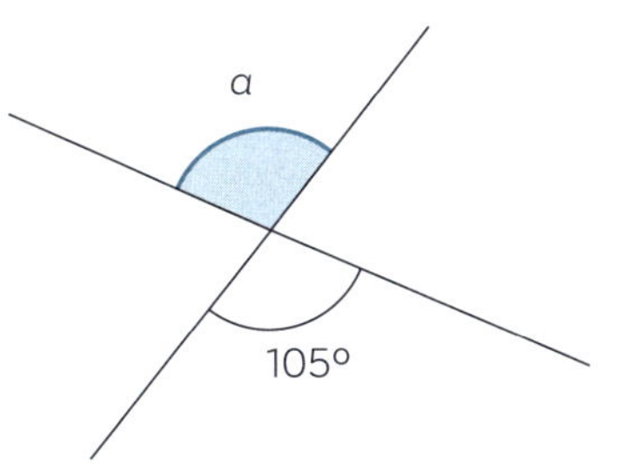

Reason(s): ______

4

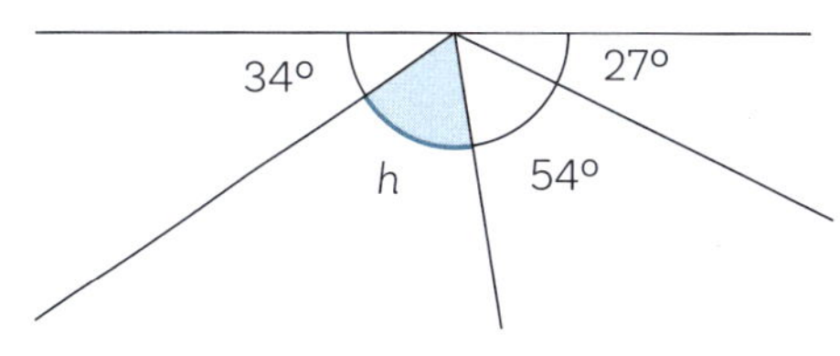

Reason(s): ______

5

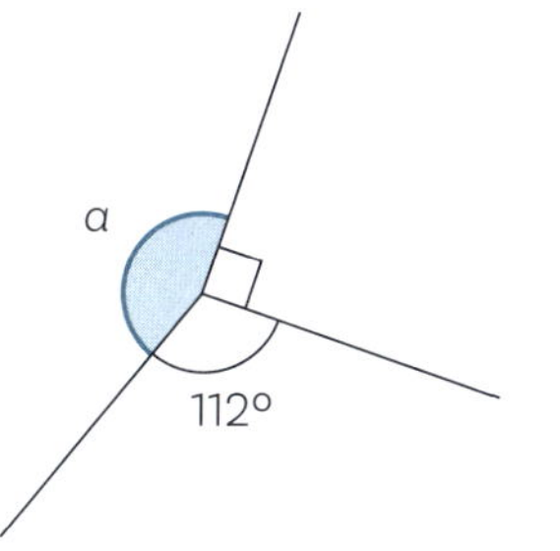

Reason(s): ______

6

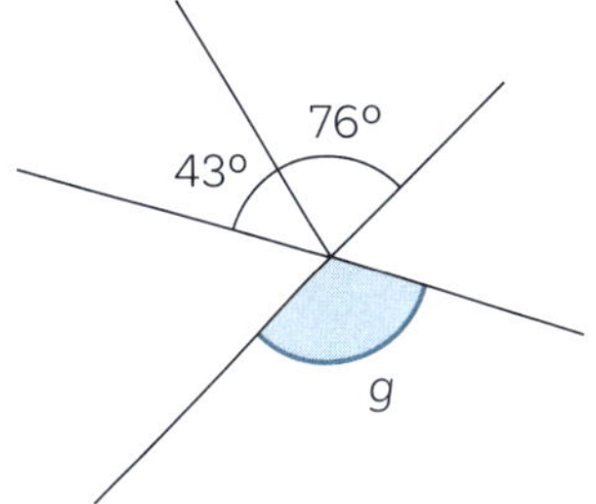

Reason(s): ______

7

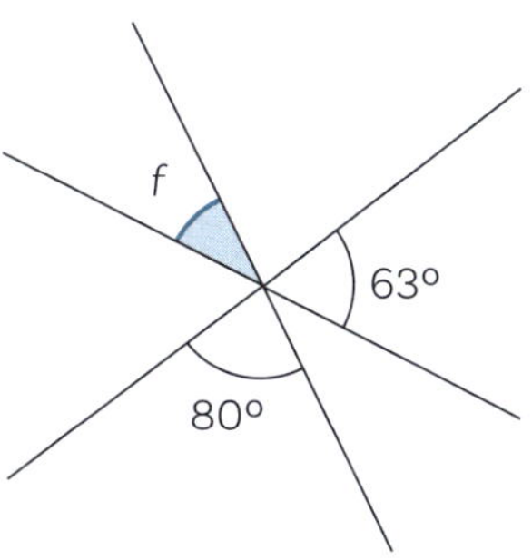

Reason(s): ______

8

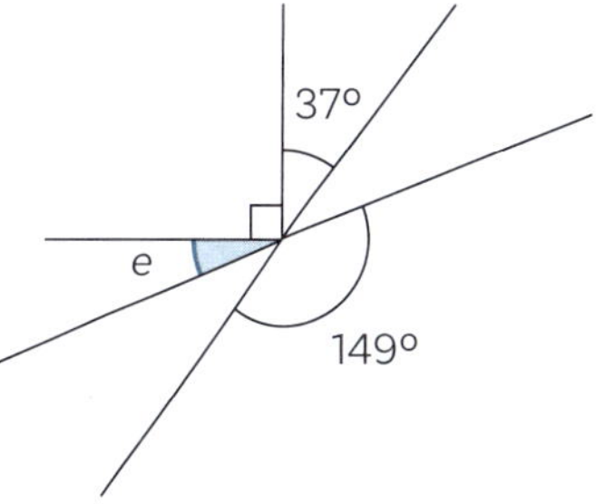

Reason(s): ______

Calculate the unknown angles and give reasons.

9

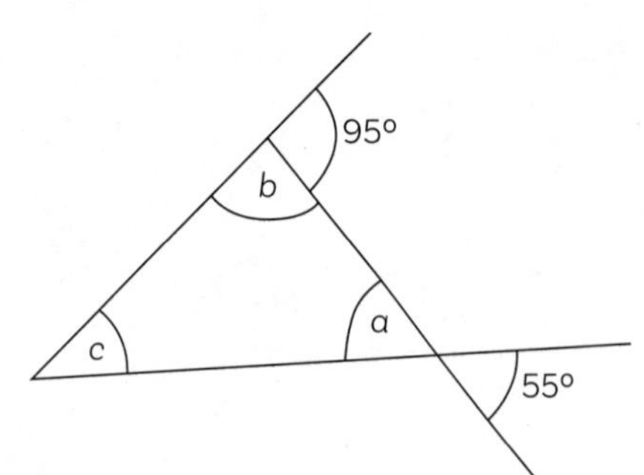

$a =$ ____ ________________________

$b =$ ____ ________________________

$c =$ ____ ________________________

10

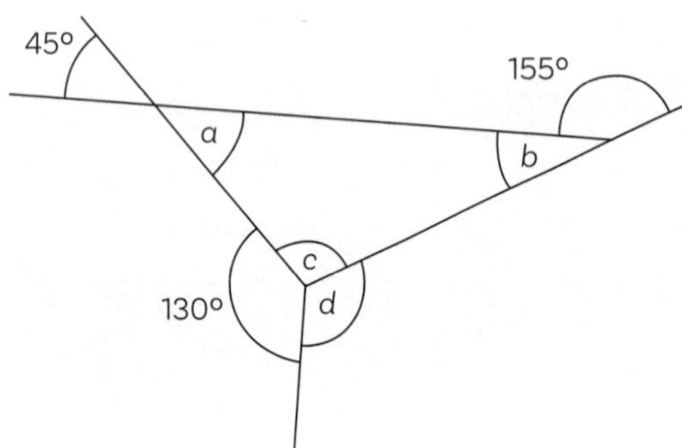

$a =$ ____ ________________________

$b =$ ____ ________________________

$c =$ ____ ________________________

$d =$ ____ ________________________

11

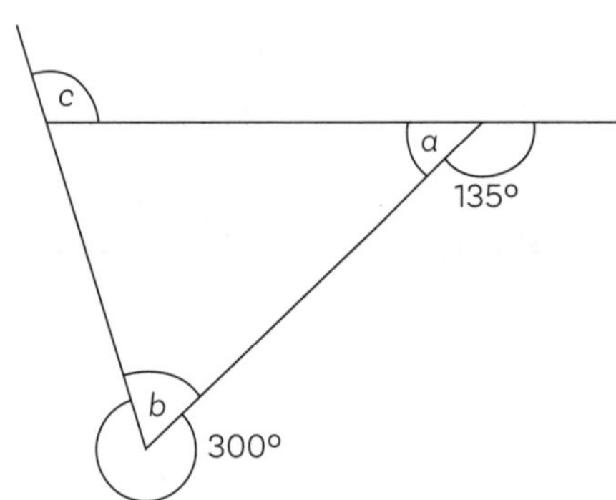

$a =$ ____ ________________________

$b =$ ____ ________________________

$c =$ ____ ________________________

12

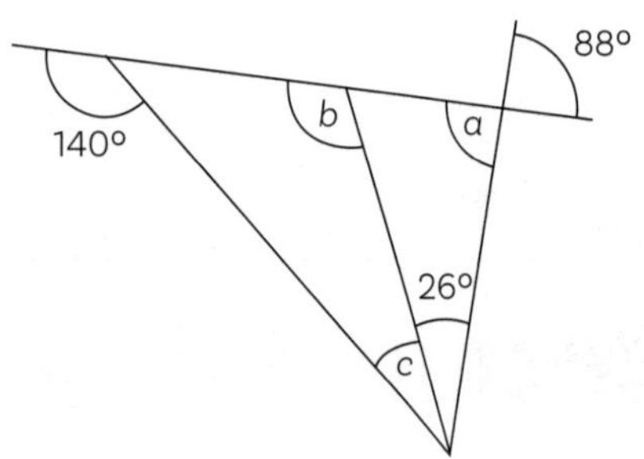

$a =$ ____ ________________________

$b =$ ____ ________________________

$c =$ ____ ________________________

13

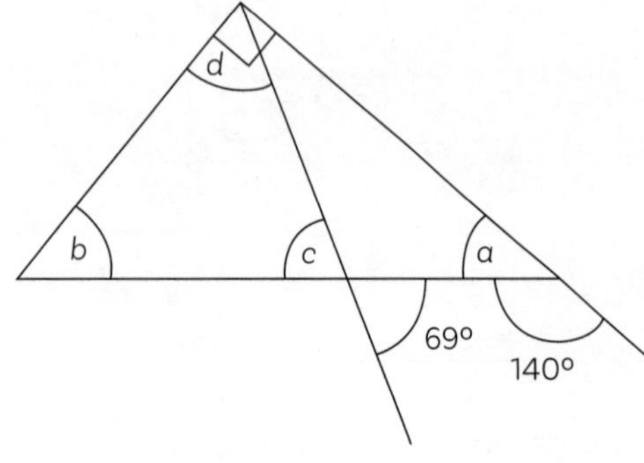

$a =$ ____ ________________________

$b =$ ____ ________________________

$c =$ ____ ________________________

$d =$ ____ ________________________

 ISBN: 9780170484084

Parallel lines

Relationship	Reason
$a = b$	Alternate angles on parallel lines are equal. (These form a '**Z**'.) (alt ∠s =, // lines)
$a = b$	Corresponding angles on parallel lines are equal. (These form an '**F**'.) (corr ∠s =, // lines)
$a + b = 180°$	Co-interior angles on parallel lines add to 180°. (These form a '**C**'.) (co-int ∠s add to 180°, // lines)

State which of these relationships applies to the following pairs of angles.

1

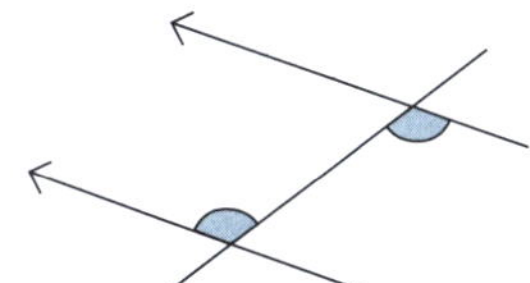

____

2

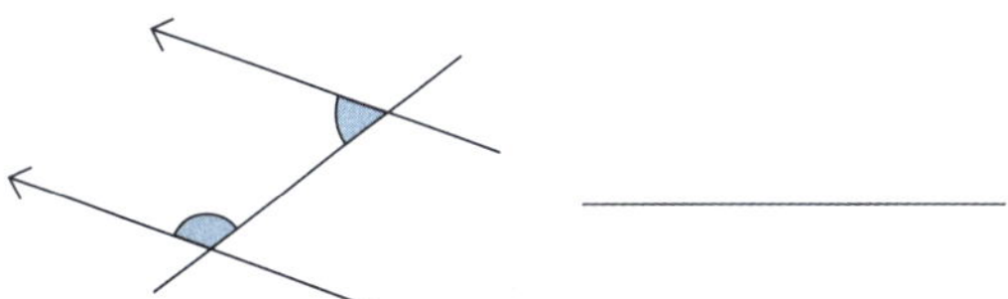

____

3

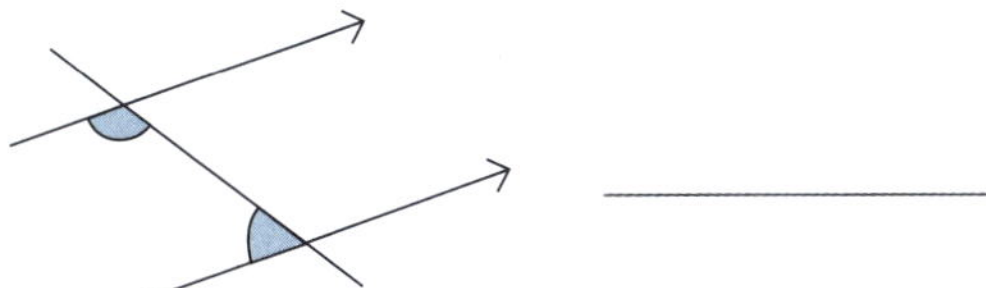

____

4

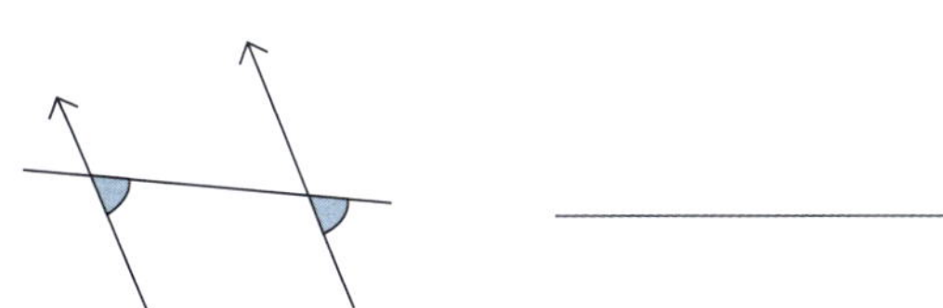

____

5

____

6

____

7

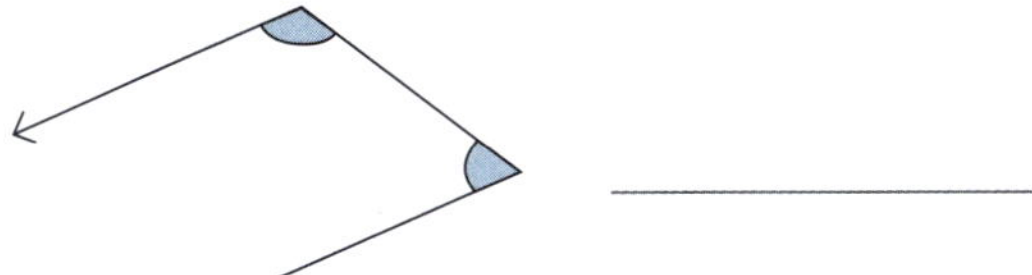

____

8

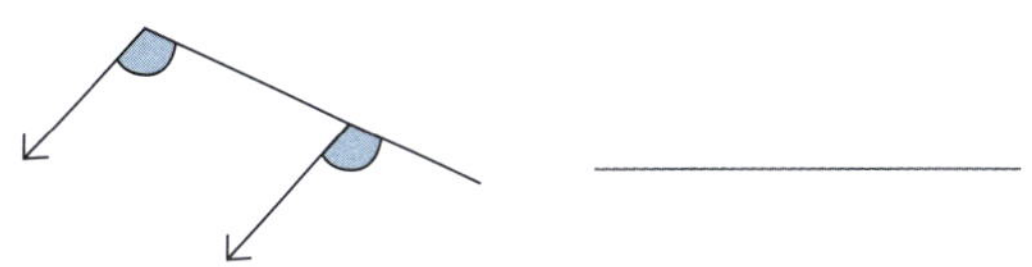

____

ISBN: 9780170484084

Calculating angles where parallel lines are involved

- Once again, calculate all the angles you can.
- Be prepared to use the basic angle rules as well as the parallel line rules.
- Sometimes you will need to give two reasons for an angle calculation.

Example: Calculate the angle θ and give reasons for each step.

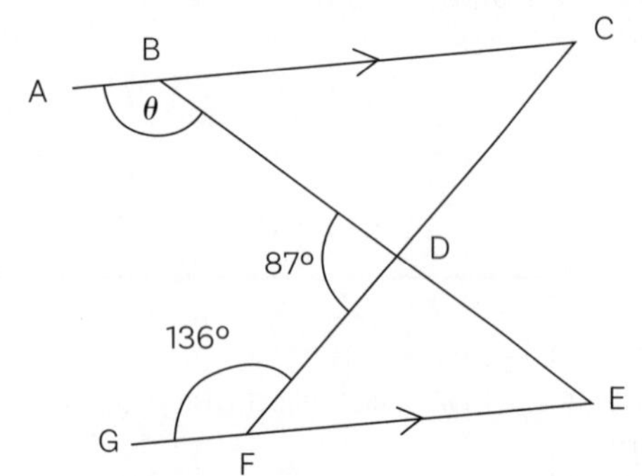

$\angle$DCB = 44° (co-int $\angle$s add to 180°, // lines)

$\angle$BDC = 93° ($\angle$s on a line = 180°)

$\therefore \angle$ABD = 137° (ext $\angle$ of Δ = sum of int opp $\angle$s)

θ = 137°

Note: there are other ways of finding the answer.

Calculate the unknown angles and give reasons for each step.

1

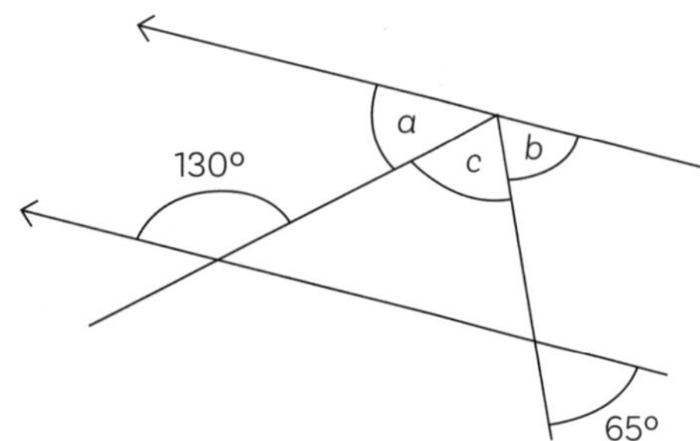

a = ____ ______________________________

b = ____ ______________________________

c = ____ ______________________________

2

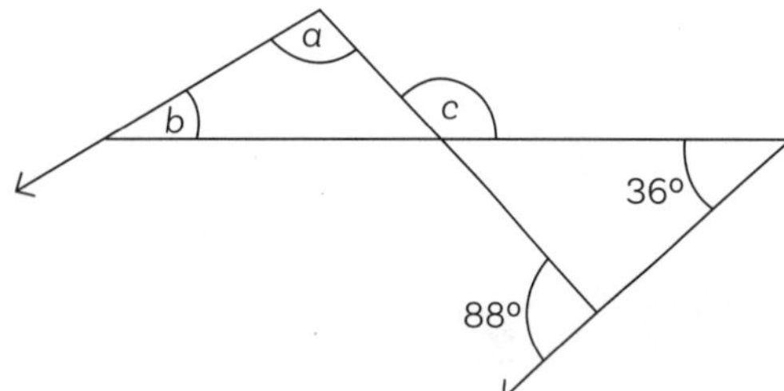

a = ____ ______________________________

b = ____ ______________________________

c = ____ ______________________________

3

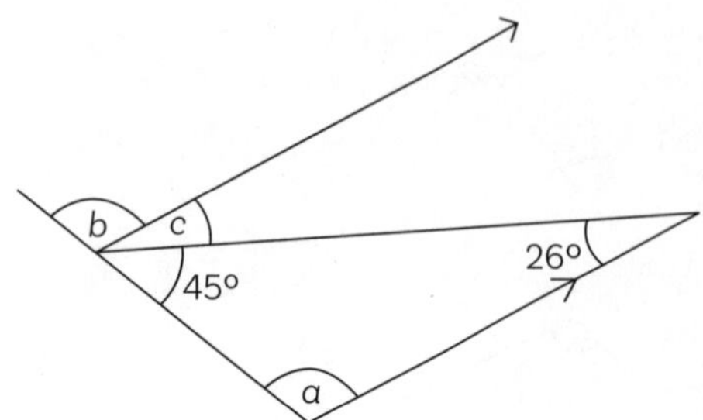

a = ____ ______________________________

b = ____ ______________________________

c = ____ ______________________________

ISBN: 9780170484084

Polygons

Triangles

- Internal angles add to 180°.

Remember:

1 For all triangles

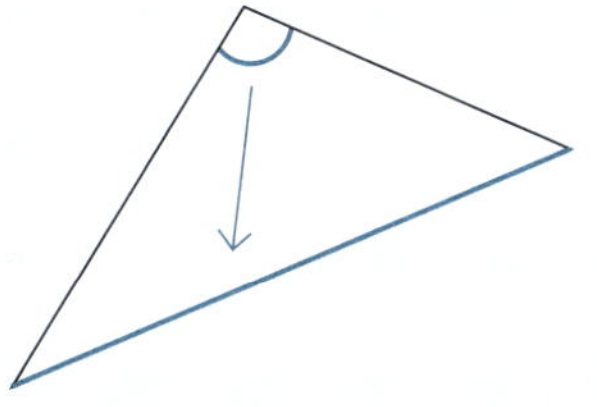

The biggest side is opposite the biggest angle.

and

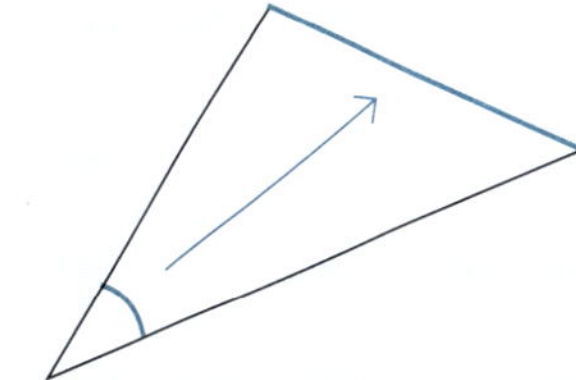

The smallest side is opposite the smallest angle.

2 Isosceles triangles

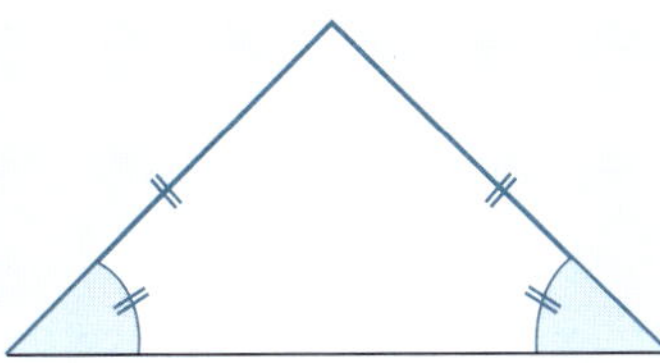

Base angles of an isosceles triangle are equal.

(isos Δ, base ∠s =)

3 Equilateral triangles

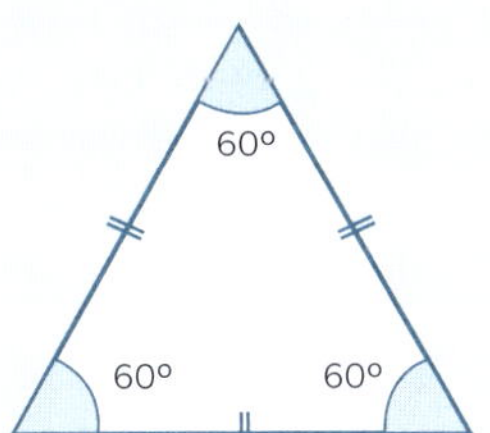

All angles = 60°.

(equilat Δ)

Calculate the unknown angles and give reasons for each step.

1

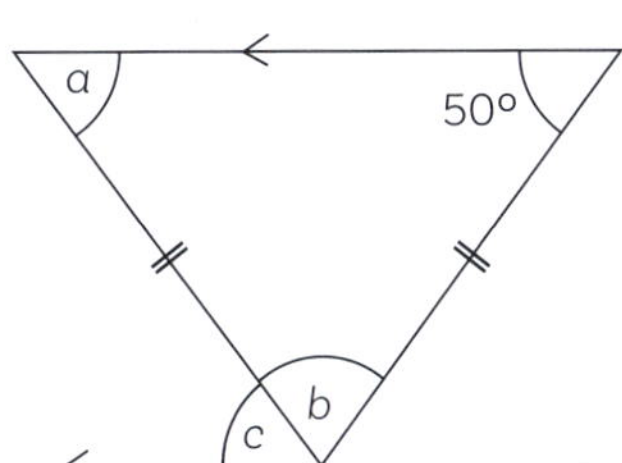

a = ____ ______________________

b = ____ ______________________

c = ____ ______________________

2

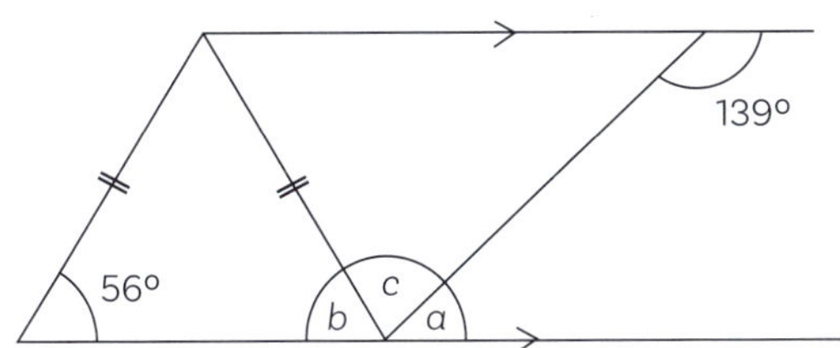

a = ____ ______________________

b = ____ ______________________

c = ____ ______________________

ISBN: 9780170484084

3

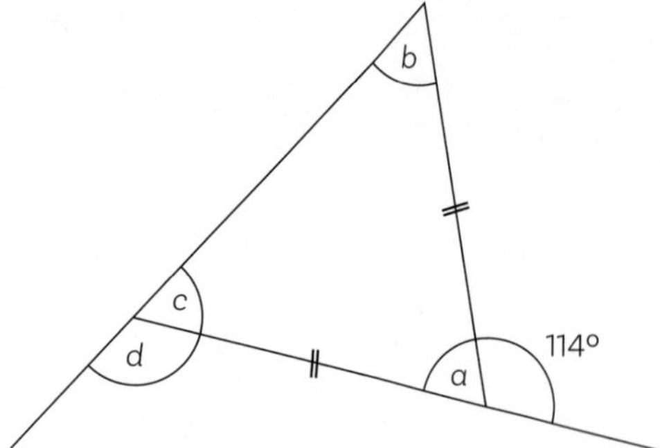

$a =$ ____ ________________________________

$b =$ ____ ________________________________

$c =$ ____ ________________________________

$d =$ ____ ________________________________

4

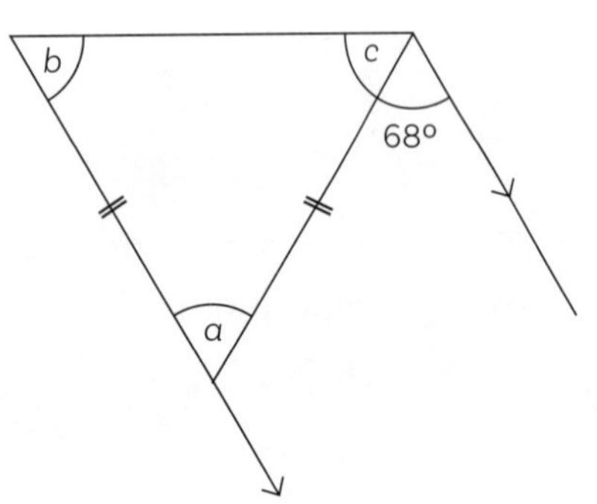

$a =$ ____ ________________________________

$b =$ ____ ________________________________

$c =$ ____ ________________________________

5

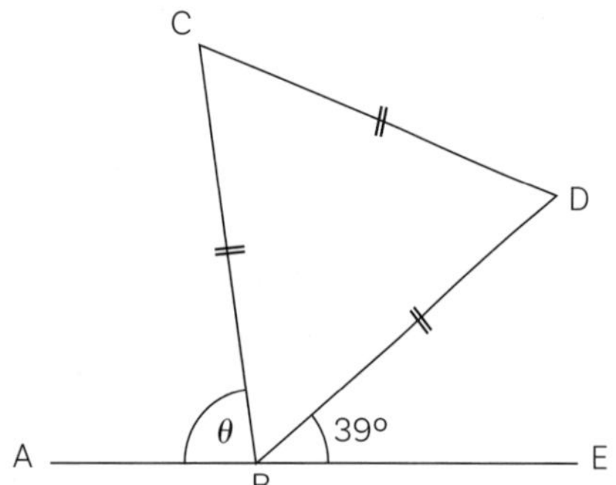

6

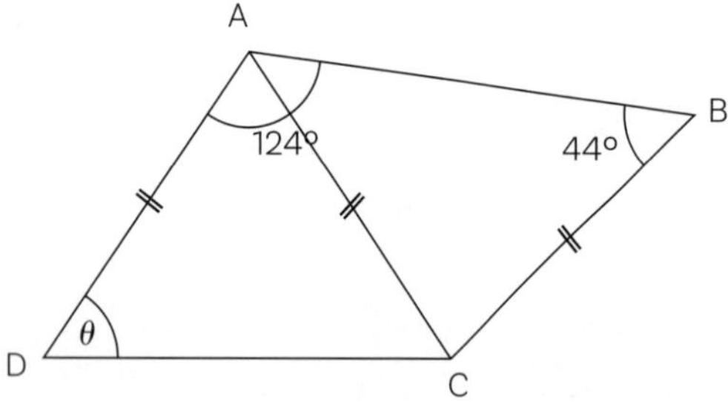

7

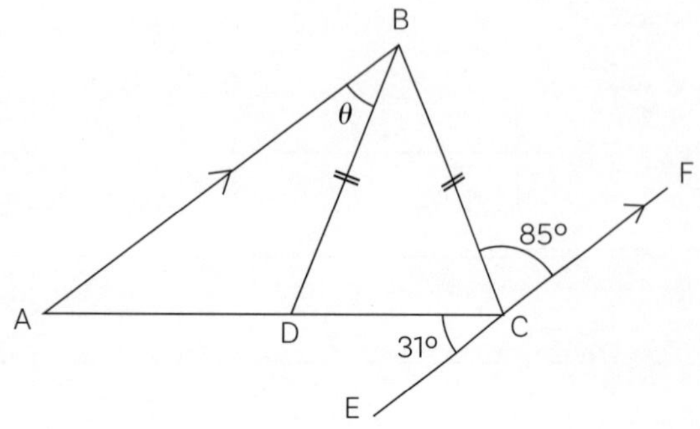

 ISBN: 9780170484084

Quadrilaterals

- Internal angles add to 360°.

Properties of special quadrilaterals

Name	Diagram	Angles	Side lengths	Parallel sides	Diagonals
Square		Right angles at each vertex	Four equal	Two pairs, opposite each other	Equal, perpendicular, bisect each other
Rhombus		No right angles, two pairs of equal angles opposite each other	Four equal	Two pairs, opposite each other	Not equal, perpendicular, bisect each other
Rectangle		Four right angles	Two equal pairs, opposite each other	Two pairs, opposite each other	Equal, not perpendicular, bisect each other
Parallelogram		No right angles, two pairs of equal angles opposite each other	Opposite sides equal, adjacent sides not	Two pairs, opposite each other	Not equal, not perpendicular, bisect each other
Kite		One pair of equal angles opposite each other, remaining angles not equal	Two pairs of equal sides, adjacent to each other	No parallel sides	Not equal, perpendicular, one is bisected by the other
Trapezium		No right angles, no equal angles	None equal	One pair parallel sides, opposite each other, remaining sides not parallel	Not equal, not perpendicular, do not bisect each other
Isosceles trapezium		Two pairs of equal angles adjacent to each other	One pair of equal sides, opposite each other, remaining sides not equal	One pair parallel sides, opposite each other, remaining sides not parallel	Equal, not perpendicular, do not bisect each other

Calculate the unknown angles and give reasons for each step.

1

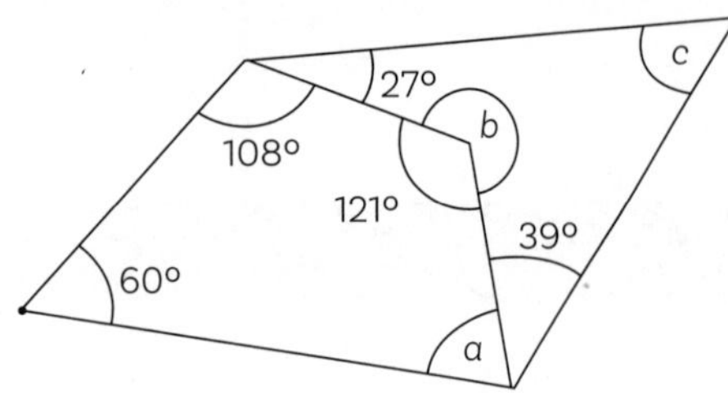

$a =$ ____ ______________________

$b =$ ____ ______________________

$c =$ ____ ______________________

2

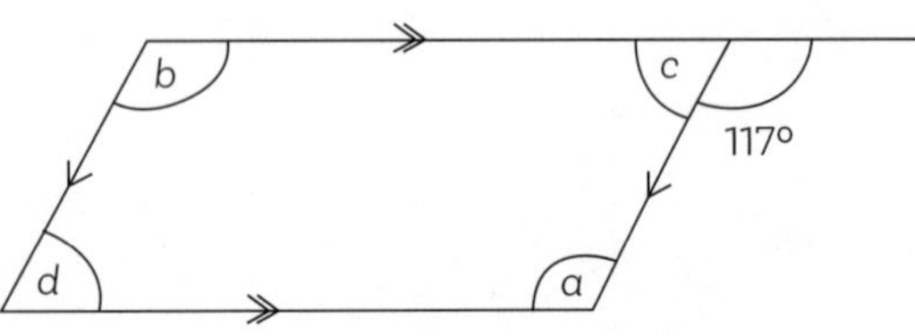

$a =$ ____ ______________________

$b =$ ____ ______________________

$c =$ ____ ______________________

$d =$ ____ ______________________

3

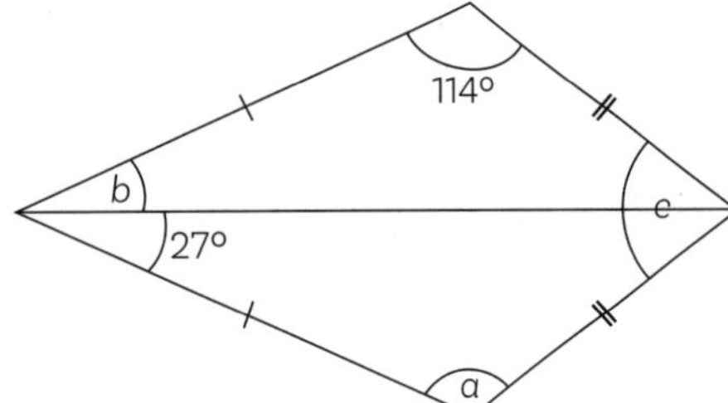

$a =$ ____ ______________________

$b =$ ____ ______________________

$c =$ ____ ______________________

4

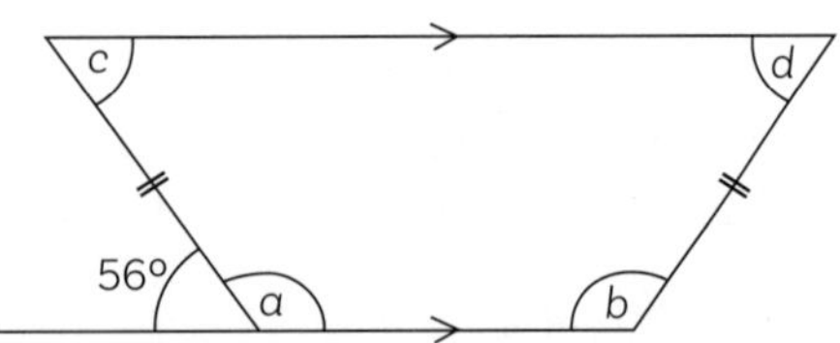

$a =$ ____ ______________________

$b =$ ____ ______________________

$c =$ ____ ______________________

$d =$ ____ ______________________

5

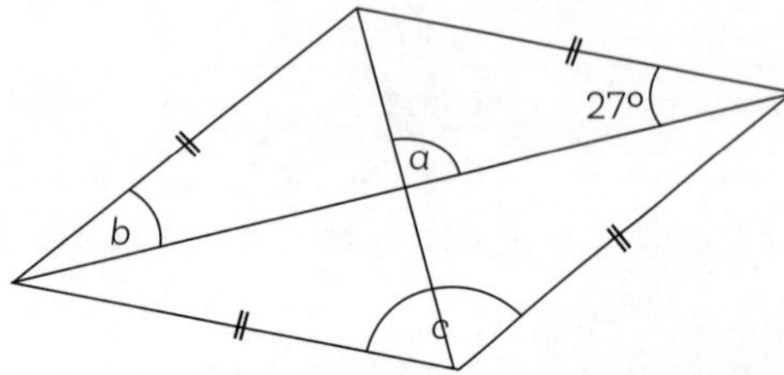

$a =$ ____ ______________________

$b =$ ____ ______________________

$c =$ ____ ______________________

ISBN: 9780170484084

Polygons in general

- Exterior and interior angles

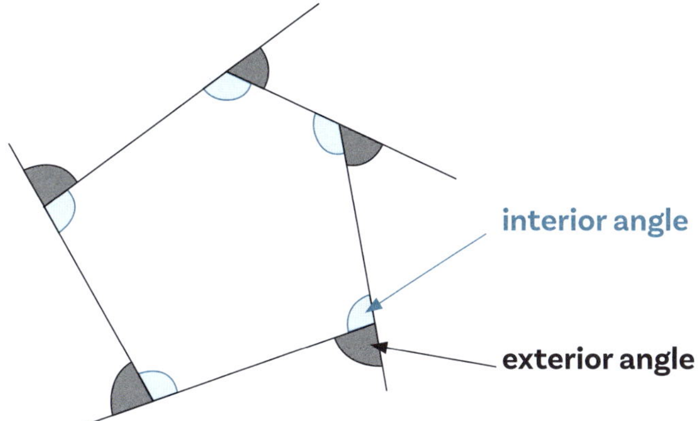

- Regular and irregular polygons

Regular polygons
– have equal sides and angles

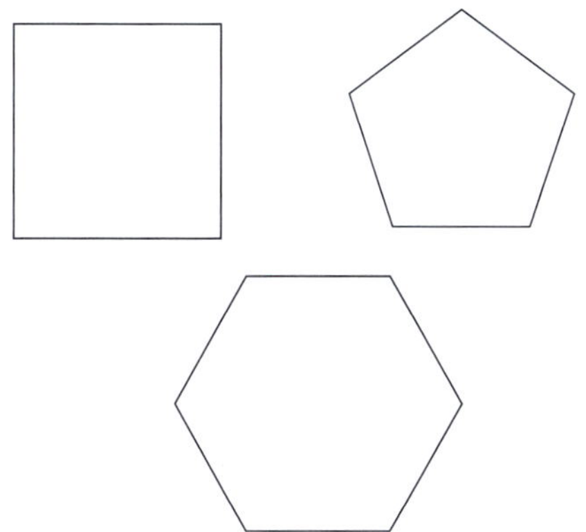

Irregular polygons
– do not have equal sides and angles

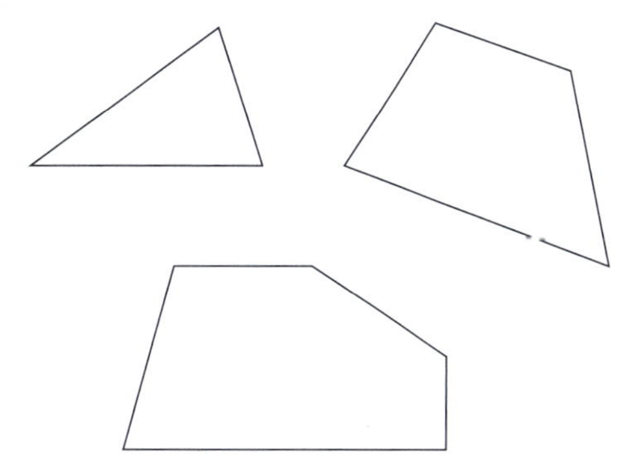

- Number of sides in polygons

Name	Number of sides
triangle	3
quadrilateral	4
pentagon	5
hexagon	6
heptagon	7
octagon	8
nonagon	9
decagon	10

ISBN: 9780170484084

Exterior angles of polygons

- These always **add to 360°** – the following diagrams demonstrate this:

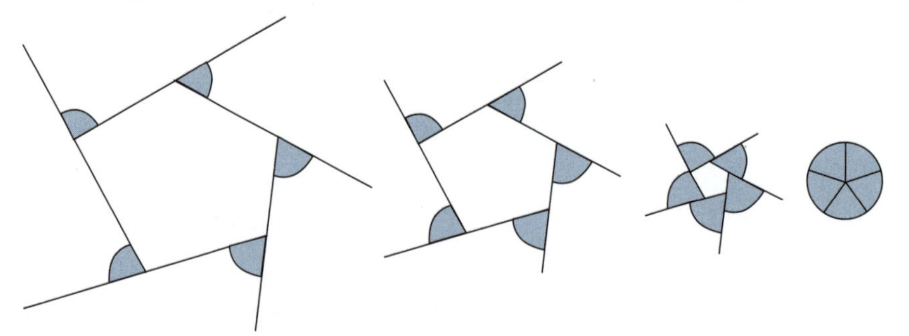

Interior angles of polygons

- Calculate the sum of the interior angles by dividing the polygon into triangles.
- You know the interior angles of a triangle add to 180°.

Name	Number of sides	Diagram	Sum of interior angles	Size of each interior angle of a *regular* polygon
Quadrilateral	**4**	1, 2	**2** x 180° = 360°	$\frac{360°}{4} = 90°$
Pentagon	**5**	1, 2, 3	**3** x 180° = 540°	$\frac{540°}{5} = 108°$
Hexagon	**6**	1, 2, 3, 4	**4** x 180° = 720°	$\frac{720°}{6} = 120°$
Heptagon	**7**	1, 2, 3, 4, 5	**5** x 180° = 900°	$\frac{900°}{7} = 128.6°$
Octagon	**8**	1, 2, 3, 4, 5, 6	**6** x 180° = 1080°	$\frac{1080°}{8} = 135°$
Any polygon	**n**	The number of triangles is always 2 fewer than the number of sides.	**(n - 2) x 180°**	$\frac{(n-2) \times 180°}{n} = 180° - \frac{360°}{n}$

ISBN: 9780170484084

Exterior and interior angles of polygons

Example 1: The diagram shows a regular octagon.

Calculate the size of θ.

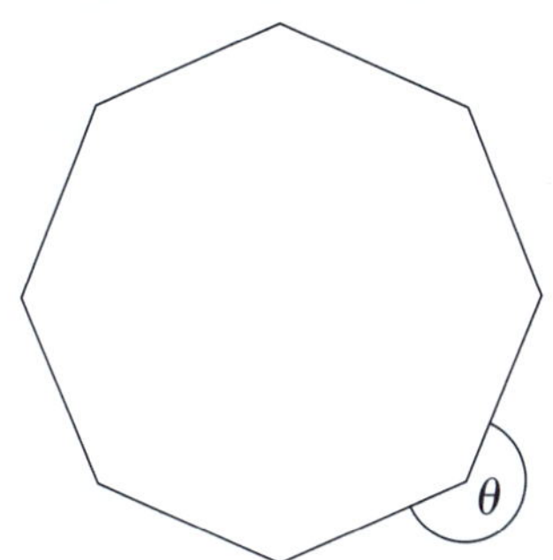

Each exterior angle = $\dfrac{360°}{8}$ = 45°

$\therefore\ \theta = 180° + 45°$

$= 225°$

Example 2: Calculate the size of θ.

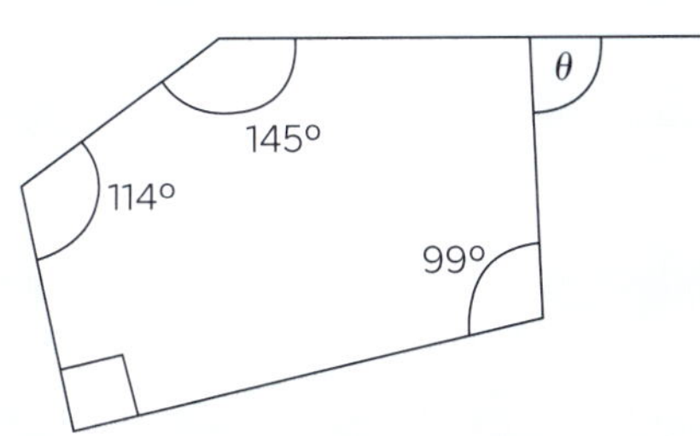

The interior angles of a pentagon add to $3 \times 180° = 540°$.

$\therefore\ 99° + 90° + 114° + 145° + (180° - \theta) = 540°$
(∠s on a line = 180°)

$\theta = 99° + 90° + 114° + 145° + 180° - 540°$

$= 88°$

Calculate the unknown angles.

1

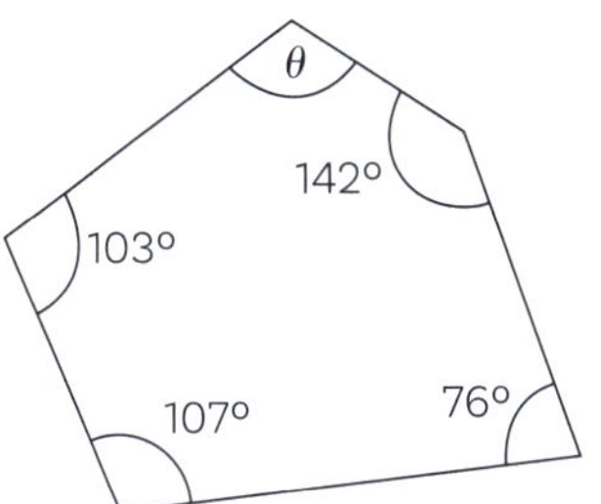

2

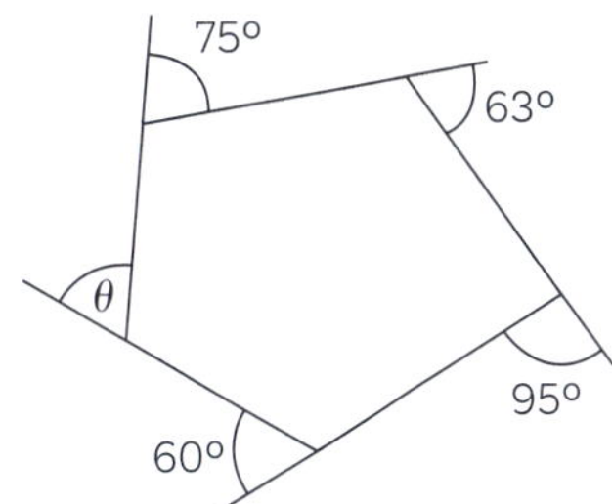

3

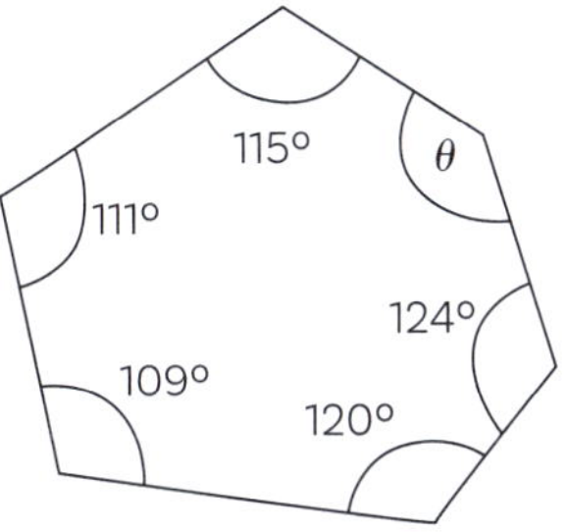

4

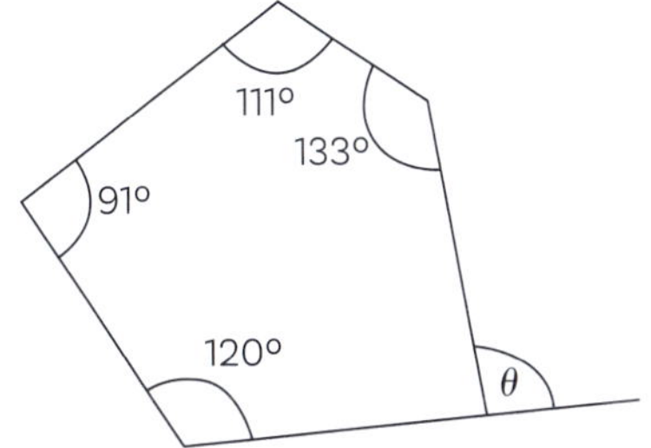

5

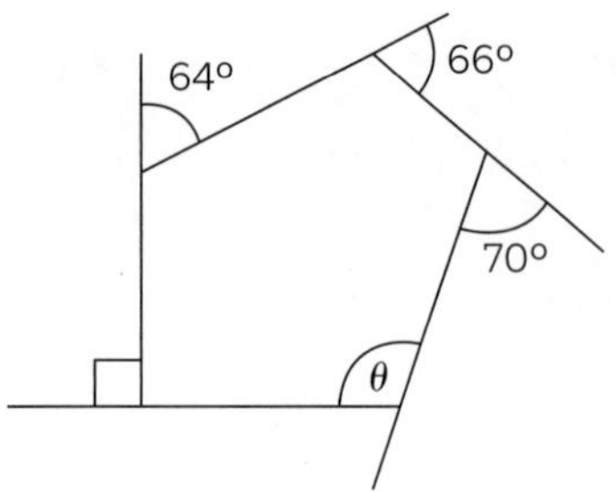

6

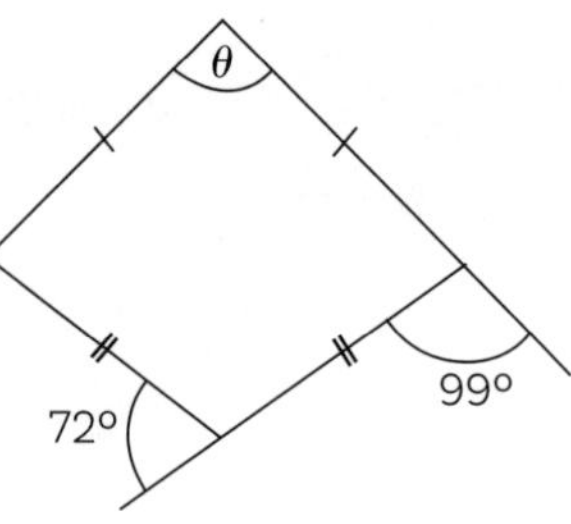

7

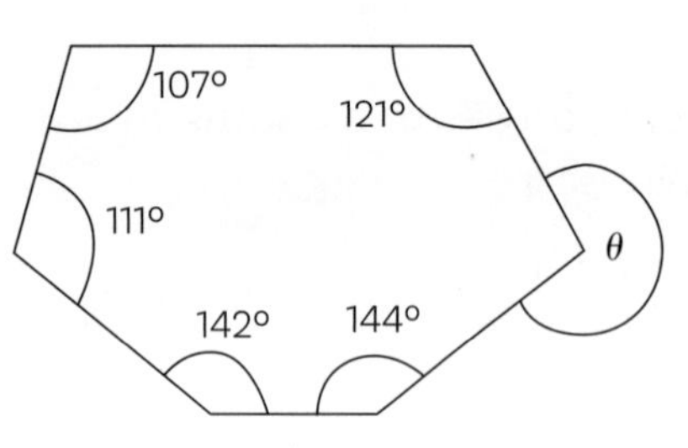

8

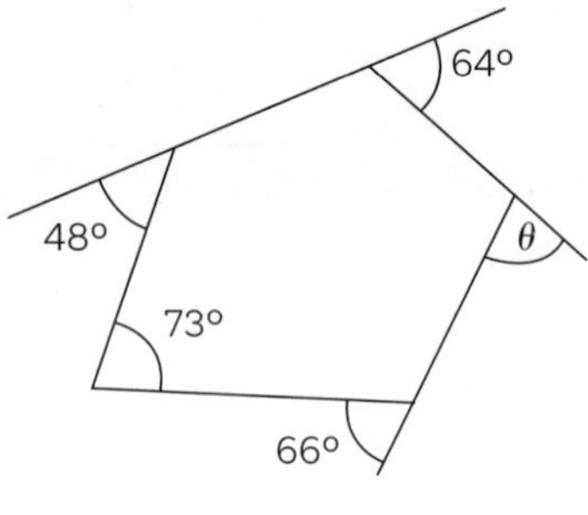

9

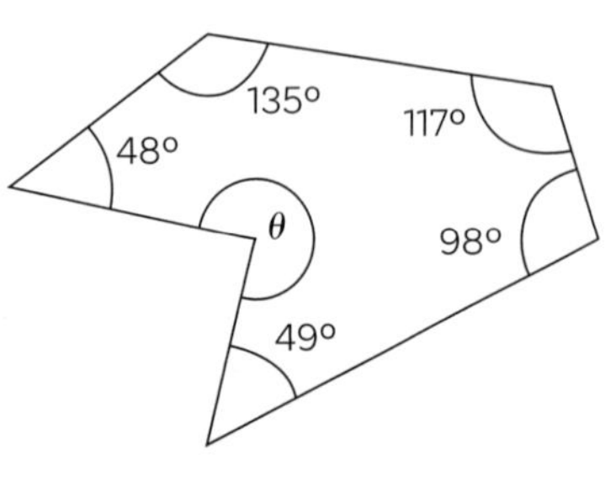

10

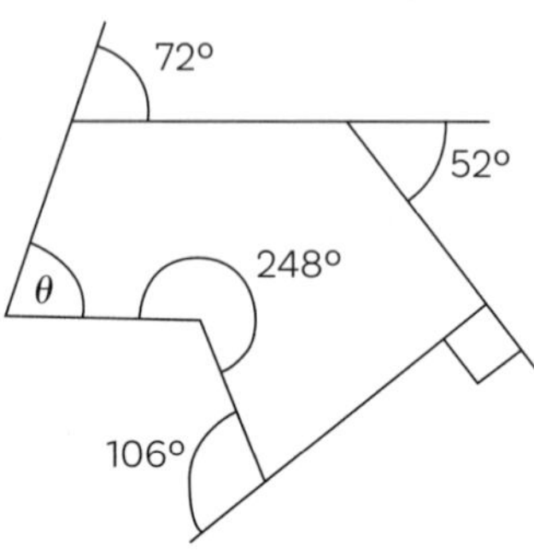

11 Calculate the size of each external angle of a regular decagon.

12 Calculate the size of each internal angle of a regular nonagon.

 ISBN: 9780170484084

The Theorem of Pythagoras

- The Theorem of Pythagoras applies to **right-angled** triangles only.
- The longest side is the **hypotenuse**.
- The hypotenuse is always **opposite** the right angle.
- The theorem is used for finding **lengths** of sides.

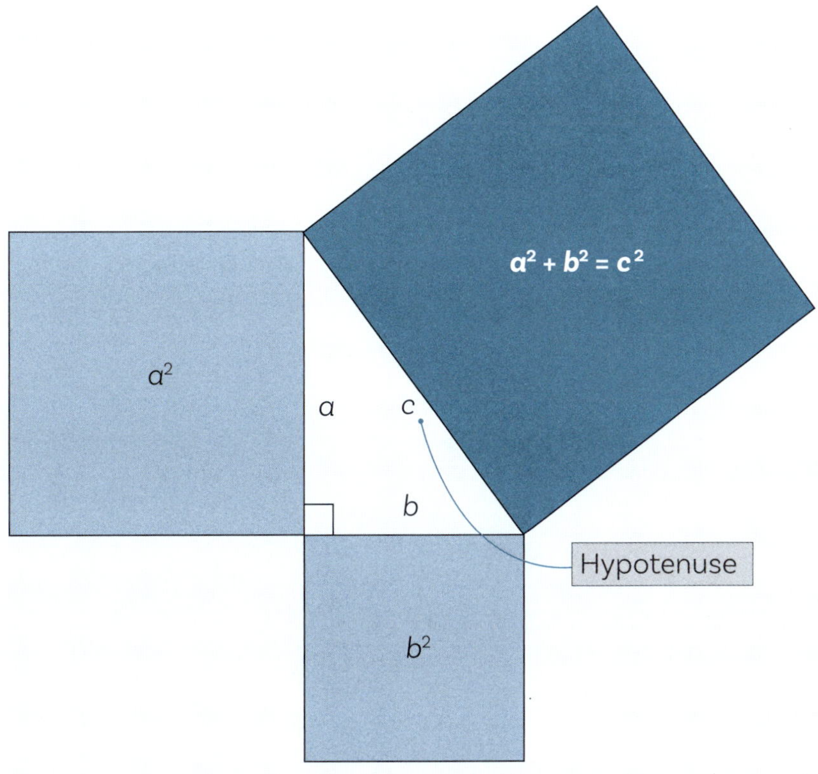

short side² + short side² = hypotenuse²

$$a^2 + b^2 = c^2$$

Finding the length of the hypotenuse

Example: Calculate the length of c.

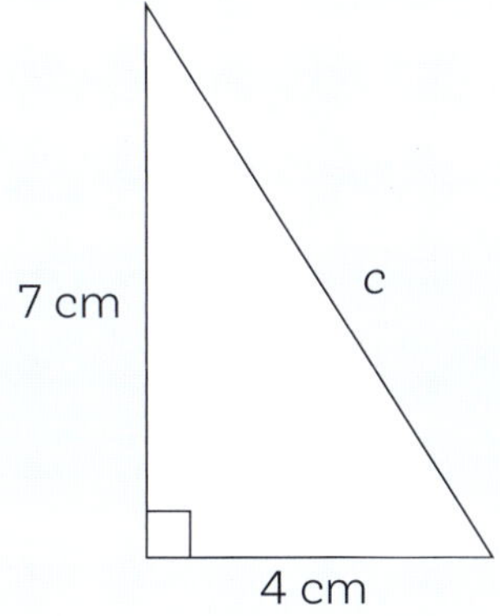

$$a^2 + b^2 = c^2$$

$$4^2 + 7^2 = c^2$$

$$\sqrt{4^2 + 7^2} = c$$

$$c = 8.06 \text{ cm (2 dp)}$$

ISBN: 9780170484084

Calculate the length of the unknown side of each triangle.

1

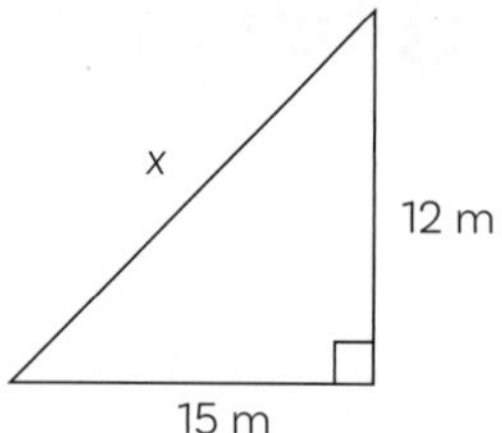

2

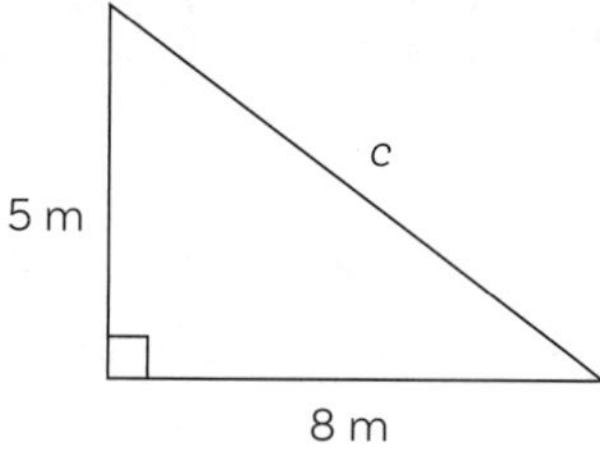

3

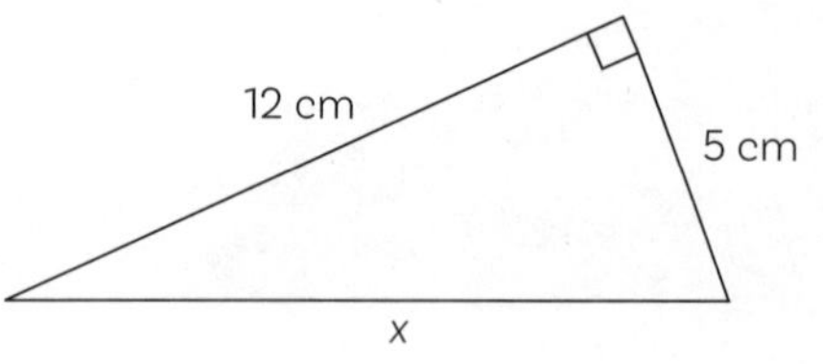

4

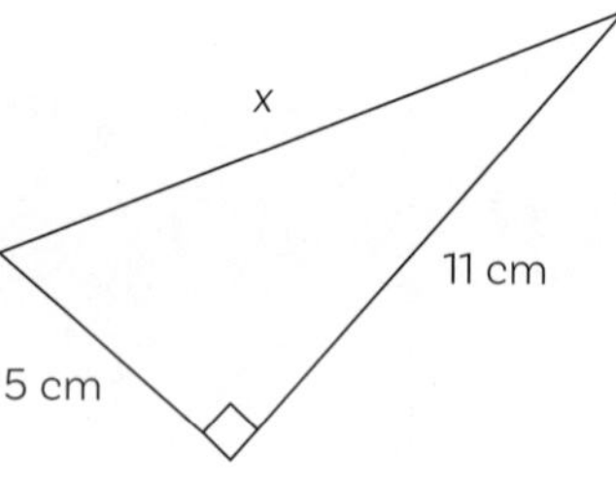

5

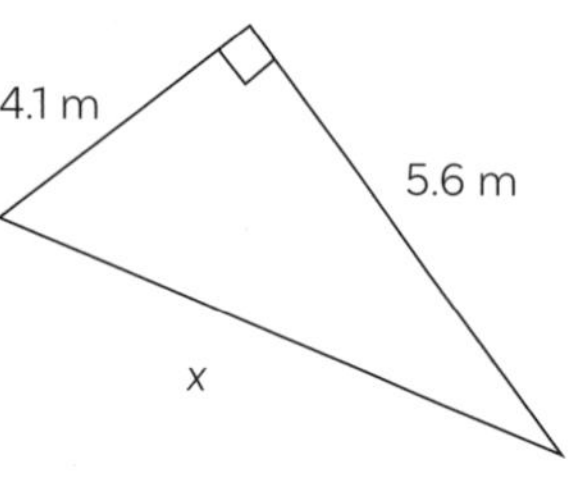

6

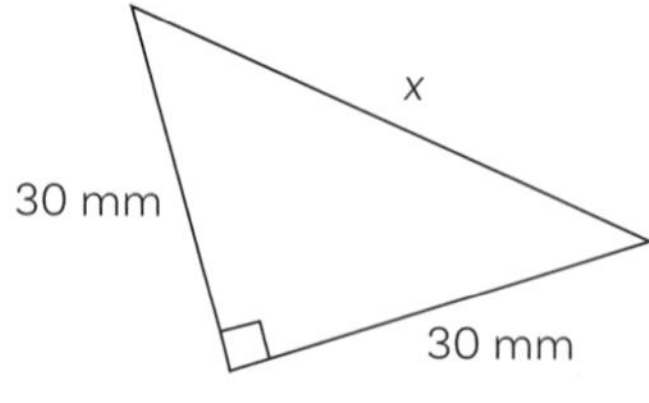

7

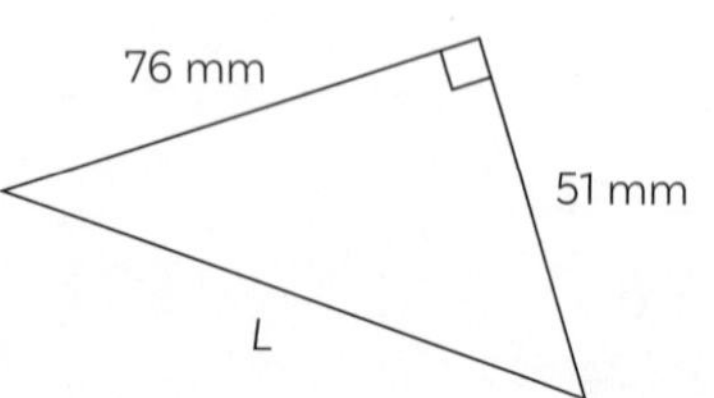

8

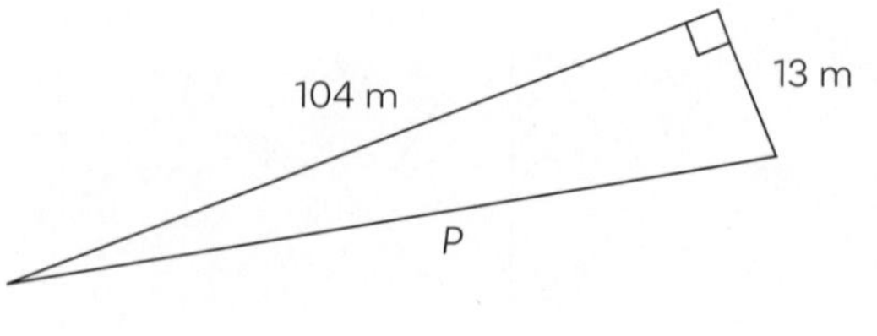

 ISBN: 9780170484084

Finding the lengths of short sides

Example: Calculate the length of y.

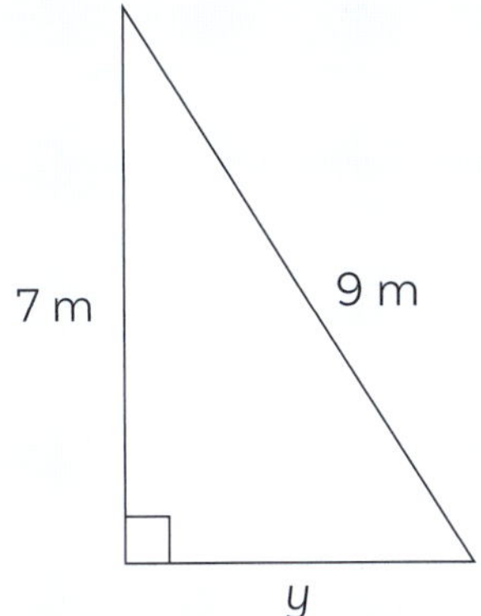

$$a^2 + b^2 = c^2$$
$$y^2 + 7^2 = 9^2$$
$$y^2 = 9^2 - 7^2$$
$$y = \sqrt{9^2 - 7^2}$$
$$y = 5.66 \text{ m (2 dp)}$$

Calculate the length of the unknown side of each triangle.

1

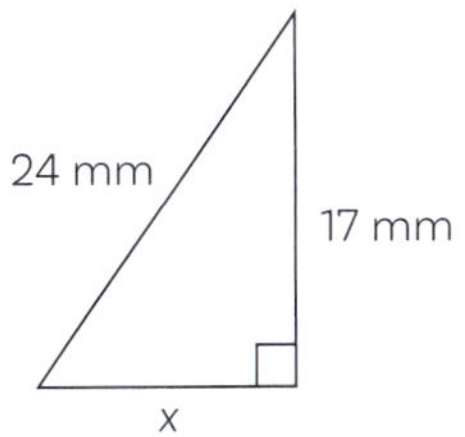

2

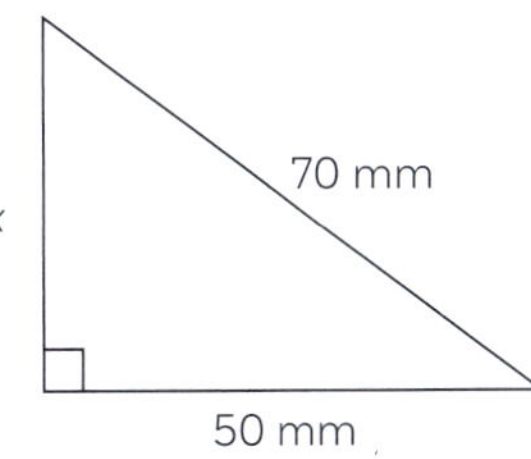

3

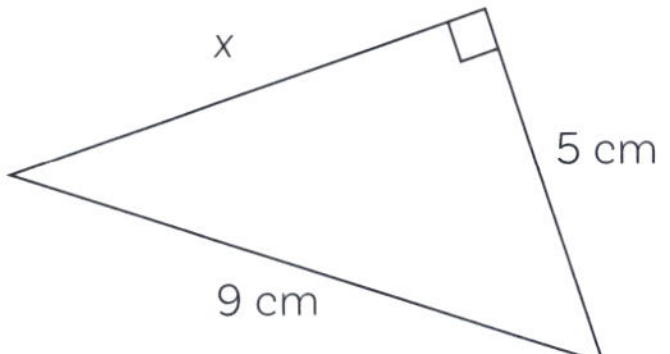

4

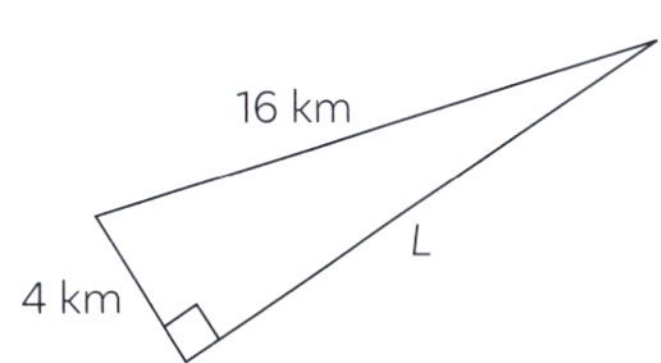

5

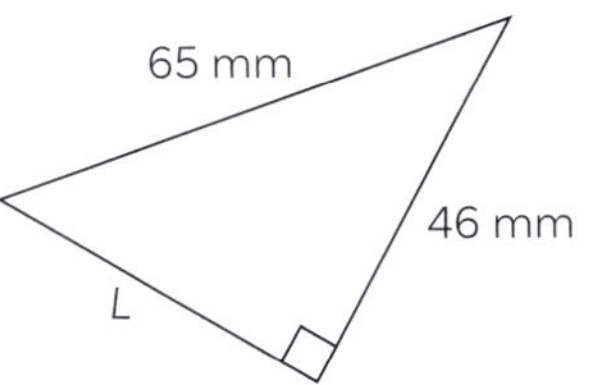

6

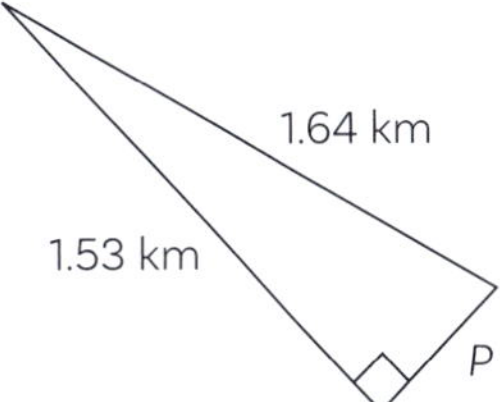

ISBN: 9780170484084

Mixing it up

Calculate the length of the unknown side of each triangle.

1

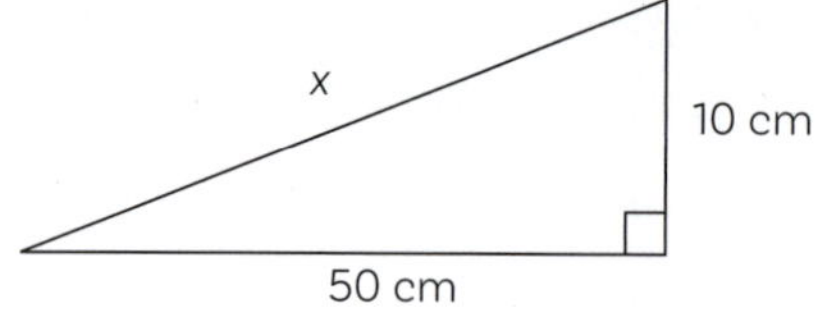

2

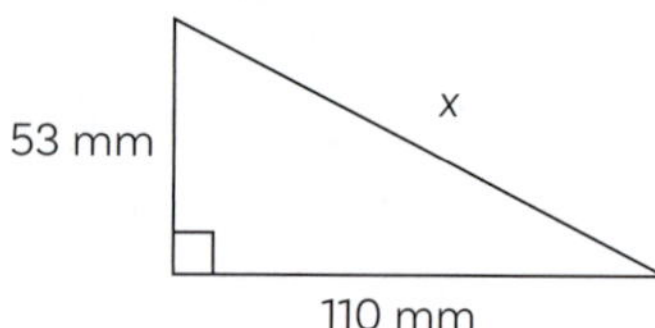

3

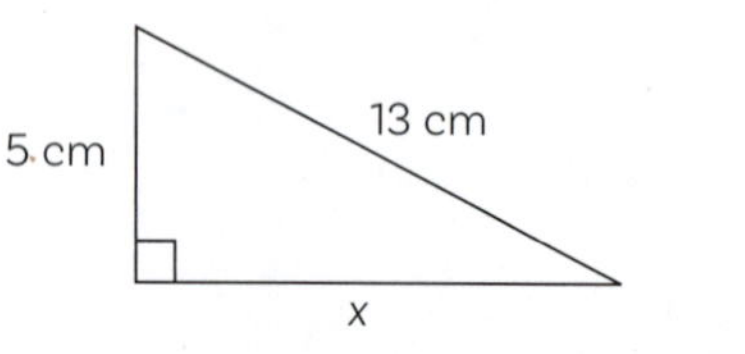

4

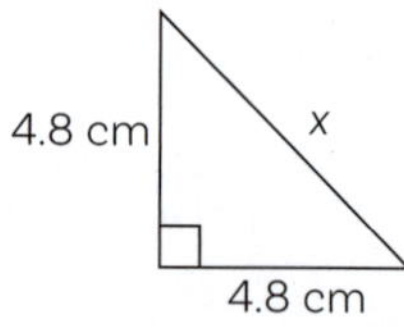

5

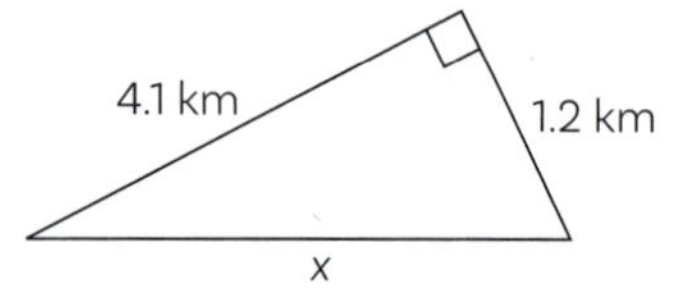

6

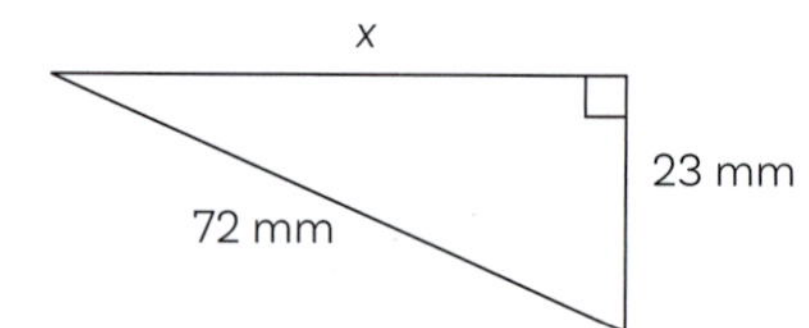

7

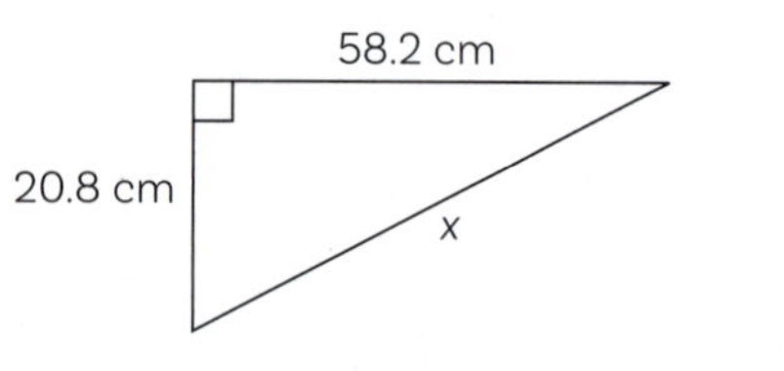

8

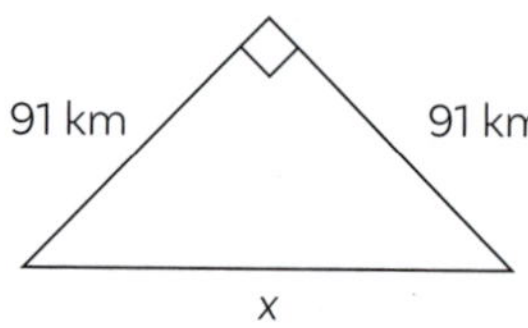

9

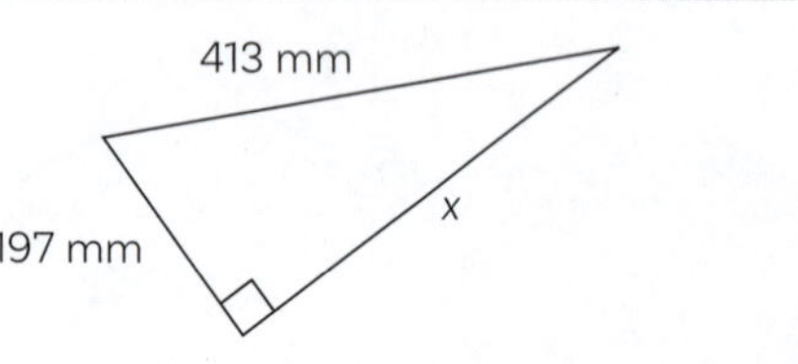

10

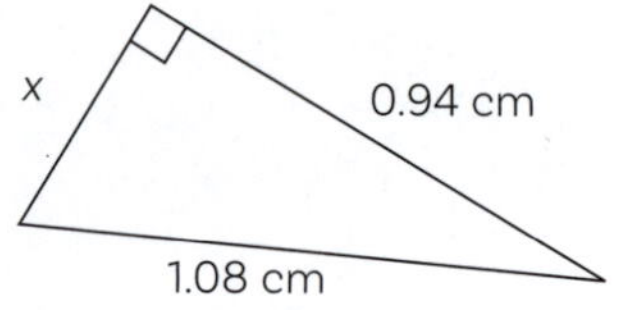

ISBN: 9780170484084

Trigonometry

Labelling sides

- Trigonometry can also be used for calculations involving **right-angled triangles**.
- However, unlike calculations using the Theorem of Pythagoras, an **angle** must be involved.

Before you begin a calculation involving trigonometry, **label** the sides of the triangle:

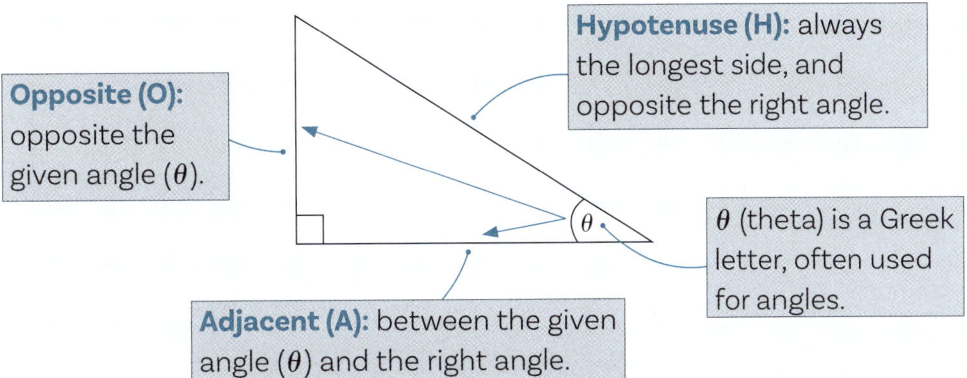

Label the following triangles with H, O and A.

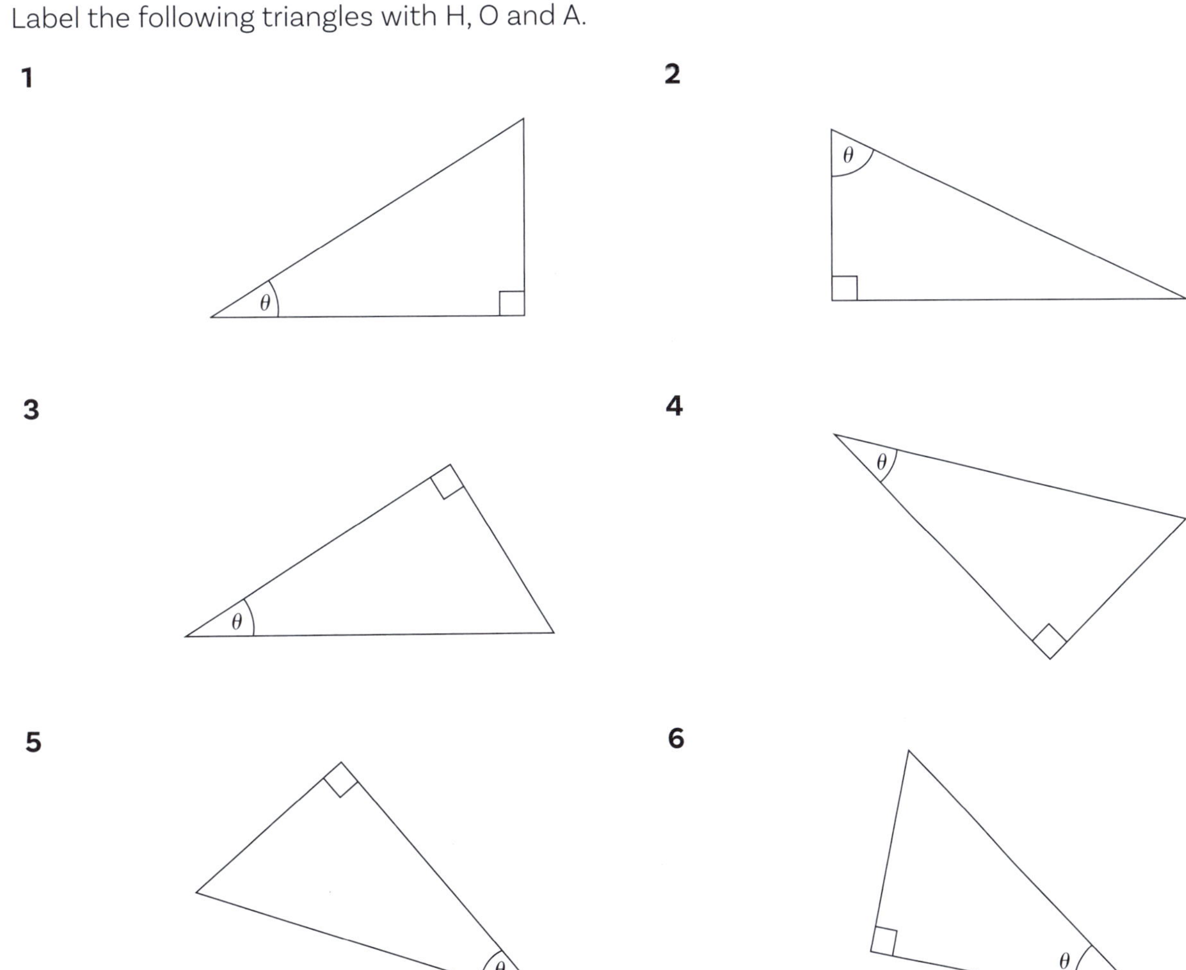

SOH CAH TOA

You will need to know about three trigonometric functions: **sin** θ (sine) **cos** θ (cosine) **tan** θ (tangent).

These are the rules in trigonometry:

$$\sin\theta = \frac{O}{H} \qquad \cos\theta = \frac{A}{H} \qquad \tan\theta = \frac{O}{A}$$

These are usually remembered as the 'word' **SOH CAH TOA**.

Organising **SOH CAH TOA** into triangles can help you work out how to use it:

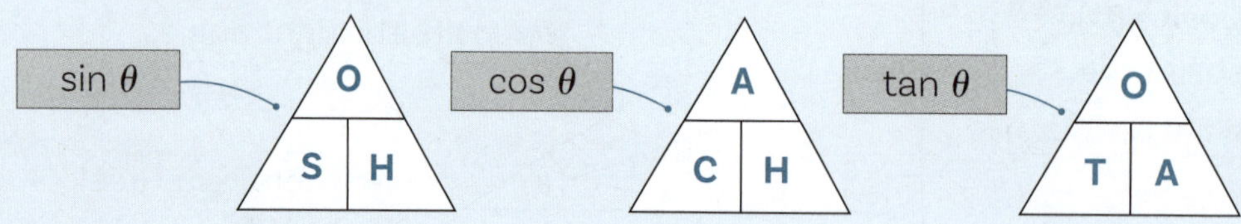

Finding short sides using sine and cosine

Example 1: Calculate the length marked x.

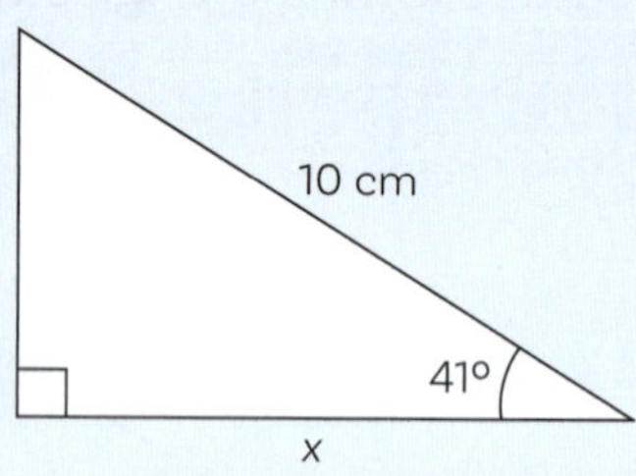

Step 1: **Label** the sides that are involved in the problem with 'A', 'O' and 'H'.

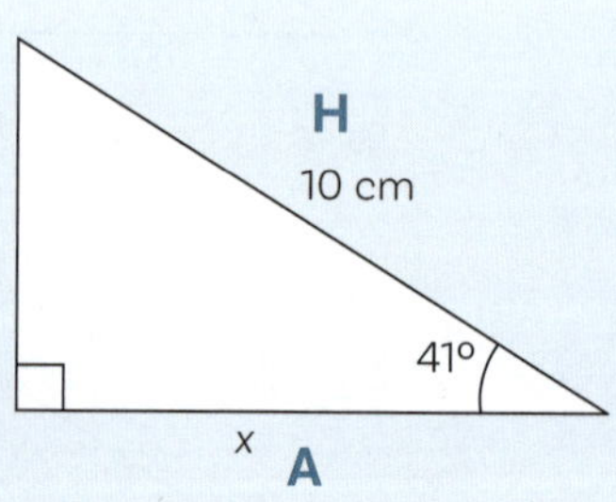

Step 2: Decide **which relationship** involves these sides. In this case we are concerned only with A and H so we must use:

$$\cos\theta = \frac{A}{H}$$

Step 3: **Substitute** the values from the triangle.

$$\cos 41° = \frac{x}{10}$$

$$x = 10\cos 41°$$

$$= 7.547 \text{ cm}$$

Check that your calculator is set to **degrees**.

Step 4: **Think about your answer – does it seem reasonable?**
In this case, the hypotenuse is 10 cm and x must be shorter, so an answer of 7.547 cm is reasonable.

ISBN: 9780170484084

Example 2: Calculate the length marked x.

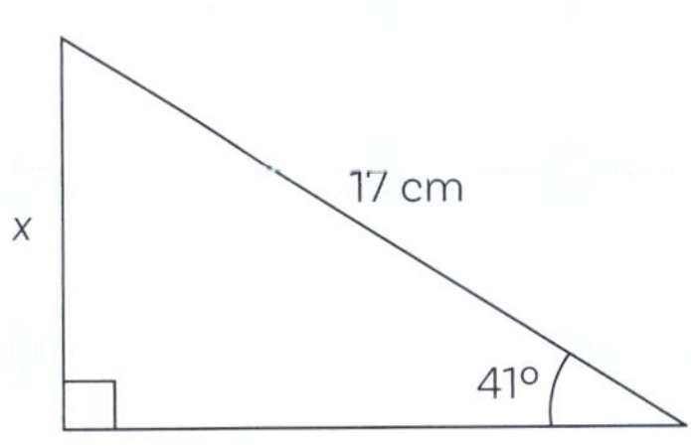

Step 1: **Label** the sides that are involved in the problem with 'A', 'O' and 'H'.

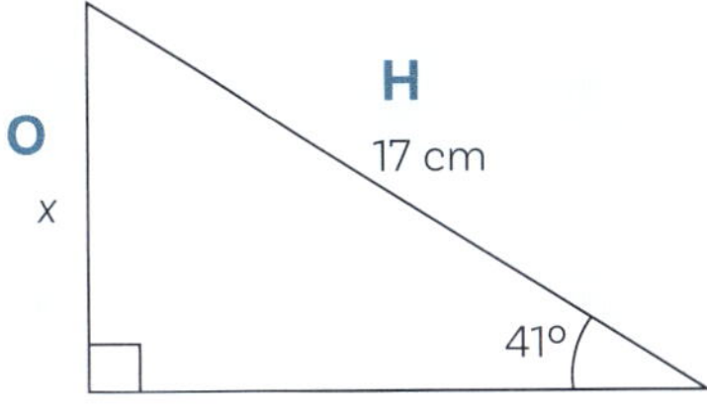

Step 2: Decide **which relationship** involves these sides. In this case we are concerned only with O and H so we must use:

$$\sin \theta = \frac{\textbf{O}}{\textbf{H}}$$

Step 3: **Substitute** the values from the triangle.

$$\sin 41^\circ = \frac{x}{17}$$

$$x = 17 \sin 41^\circ$$

$$= 11.15 \text{ cm}$$

Step 4: **Think about your answer — does it seem reasonable?**
In this case, the hypotenuse is 17 cm and *x* must be shorter, so an answer of 11.15 cm is reasonable.

Calculate the length of the unknown side of each triangle.

1

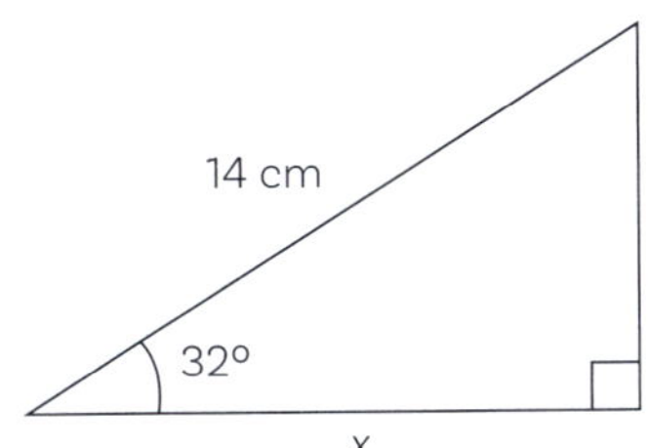

2

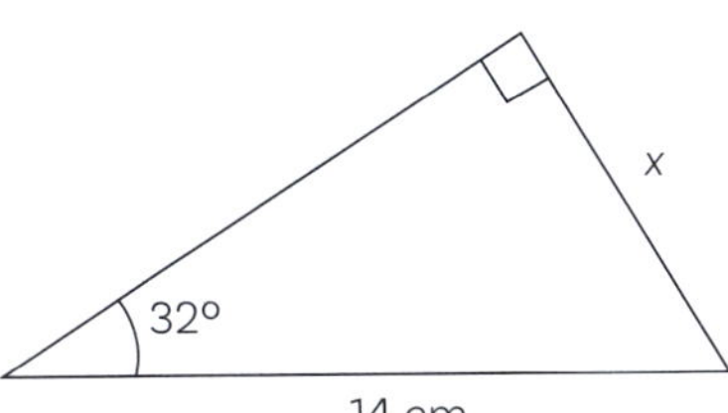

3

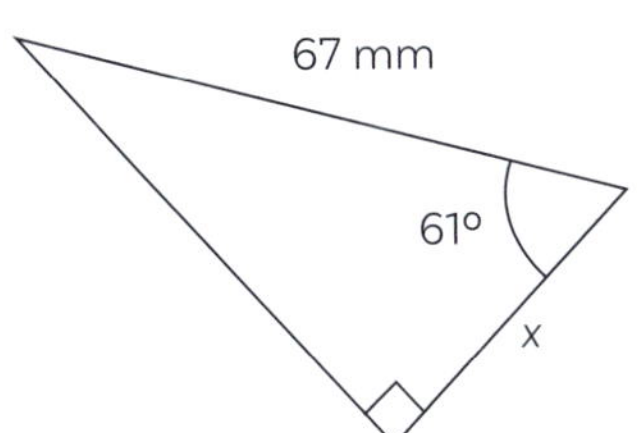

4

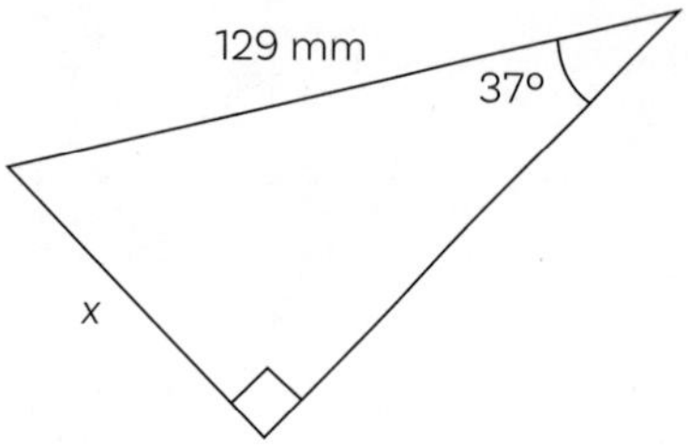

129 mm
37°
x

5

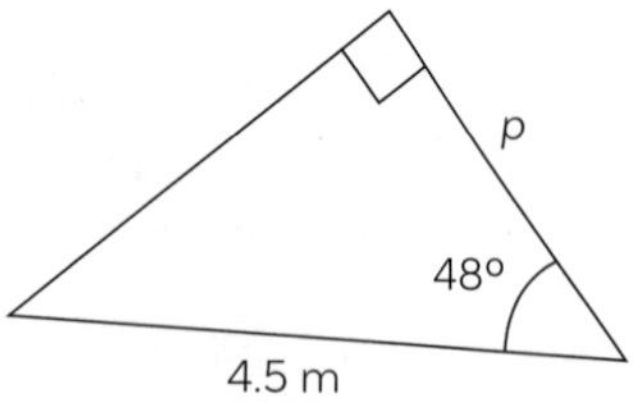

p
48°
4.5 m

6

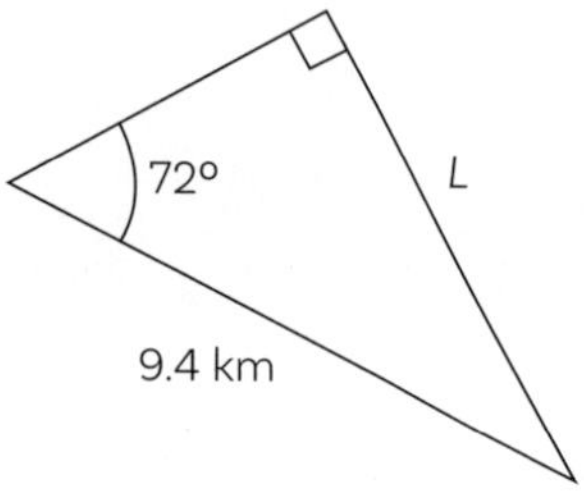

72°
L
9.4 km

7

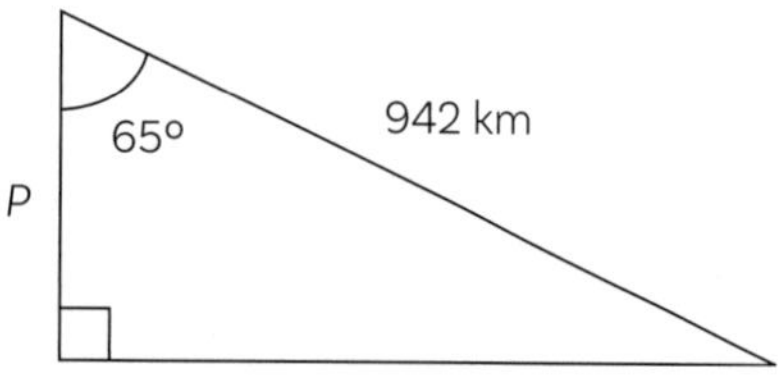

65°
942 km
P

8

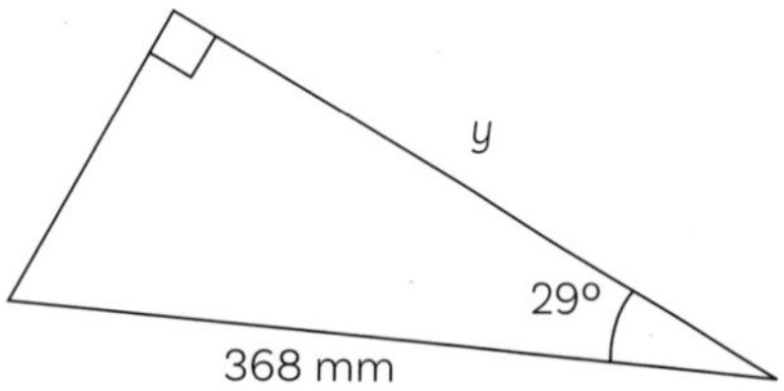

y
29°
368 mm

9

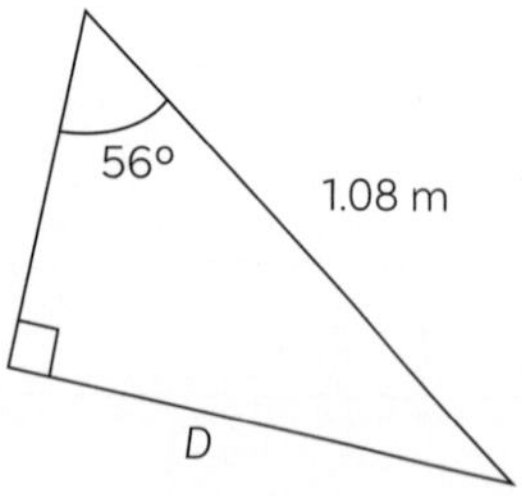

56°
1.08 m
D

10

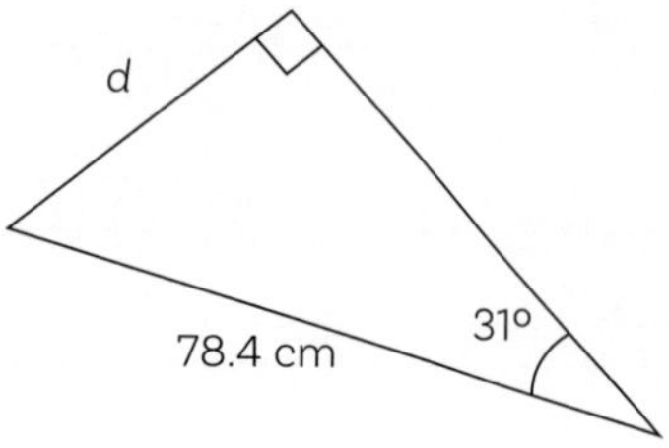

d
31°
78.4 cm

 ISBN: 9780170484084

Finding long sides (hypotenuses) using sine and cosine

This is the same process, but the algebra is a little different.

Example 1: Calculate the length marked *x*.

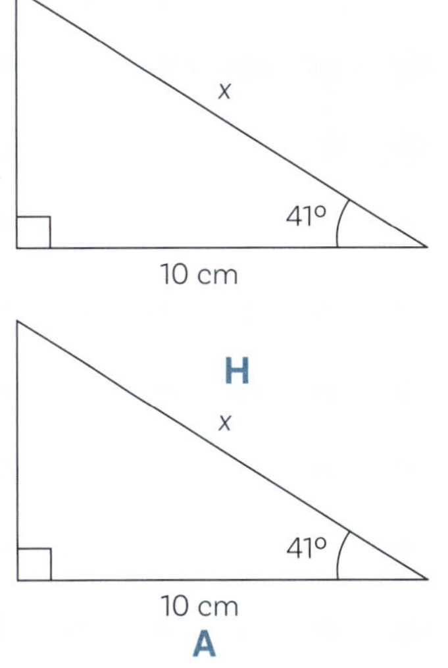

Step 1: **Label** the sides that are involved in the problem with 'A', 'O' and 'H'.

Step 2: Decide **which relationship** involves these sides. In this case we are concerned only with A and H so we must use:

$$\cos\theta = \frac{\mathbf{A}}{\mathbf{H}}$$

Step 3: **Substitute** the values from the triangle.

$$\cos 41^\circ = \frac{10}{x}$$

$$x\cos 41^\circ = 10$$

$$x = \frac{10}{\cos 41^\circ}$$

$$x = 13.25\text{ cm}$$

Step 4: **Think about your answer – does it seem reasonable?**
In this case, the adjacent side is 10 cm and *x* must be longer, so an answer of 13.25 cm is reasonable.

Example 2: Calculate the length marked *x*.

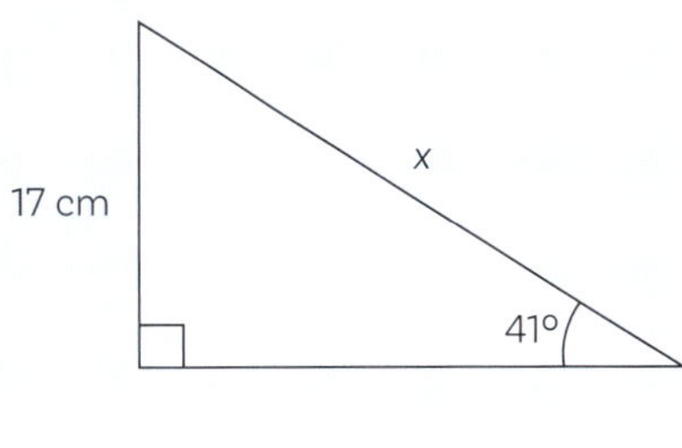

Step 1: **Label** the sides that are involved in the problem with 'A', 'O' and 'H'.

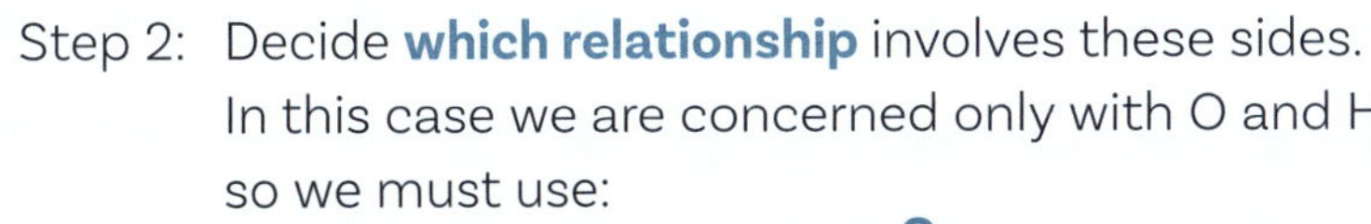

Step 2: Decide **which relationship** involves these sides. In this case we are concerned only with O and H so we must use:

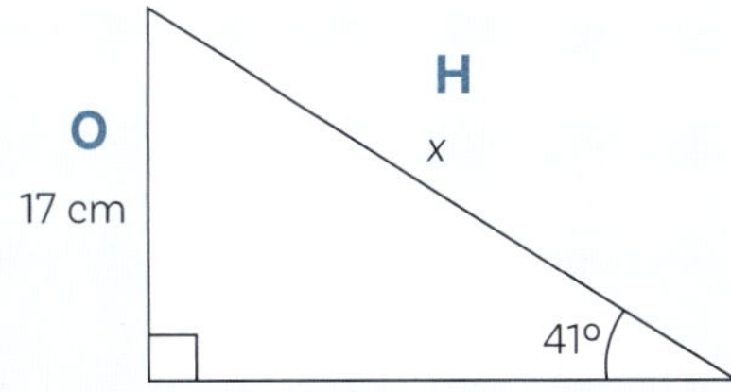

$$\sin\theta = \frac{\mathbf{O}}{\mathbf{H}}$$

Step 3: **Substitute** the values from the triangle.

$$\sin 41^\circ = \frac{17}{x}$$

$$x\sin 41^\circ = 17$$

$$x = \frac{17}{\sin 41^\circ}$$

$$x = 25.91\text{ cm}$$

Step 4: **Think about your answer – does it seem reasonable?**
In this case, the opposite side is 17 cm and *x* must be longer, so an answer of 25.91 cm is reasonable.

Calculate the length of the unknown side of each triangle.

1

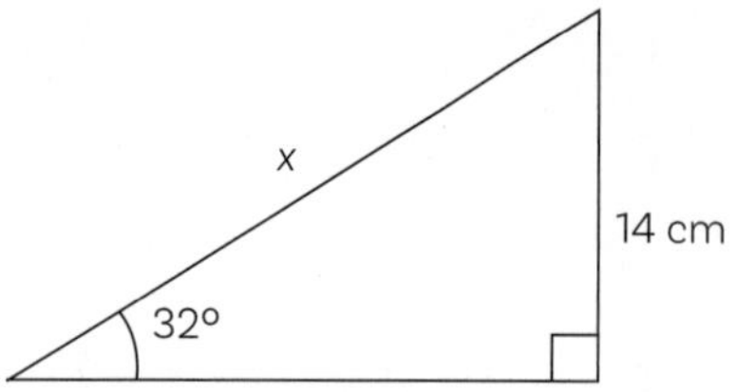

x

14 cm

32°

2

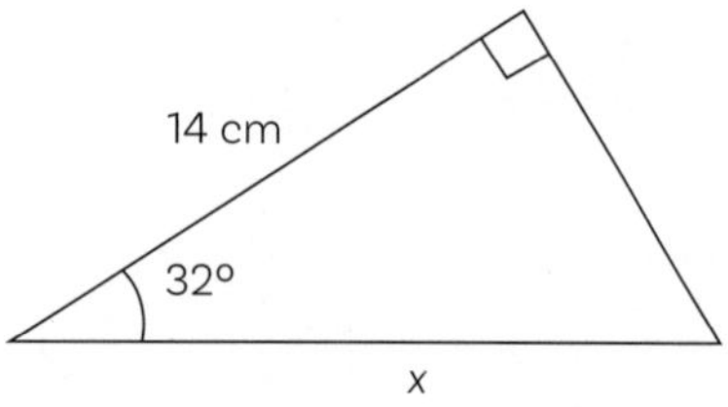

14 cm

32°

x

3

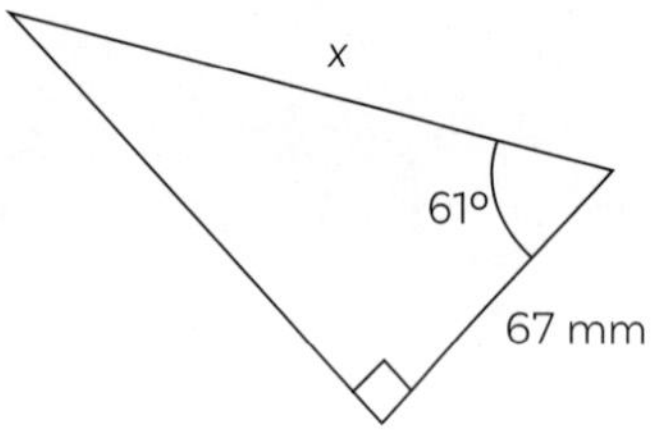

x

61°

67 mm

4

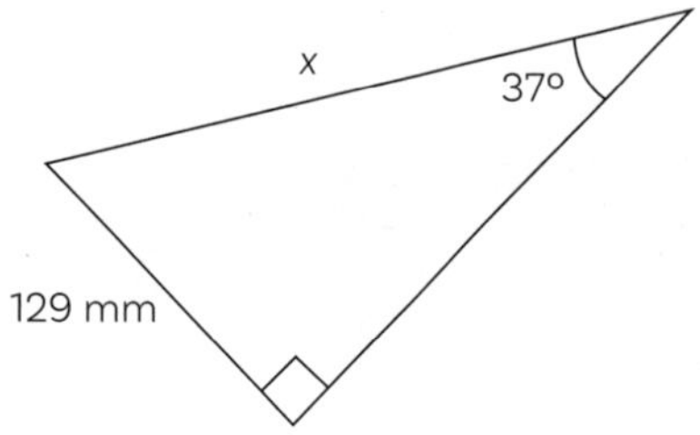

x

37°

129 mm

5

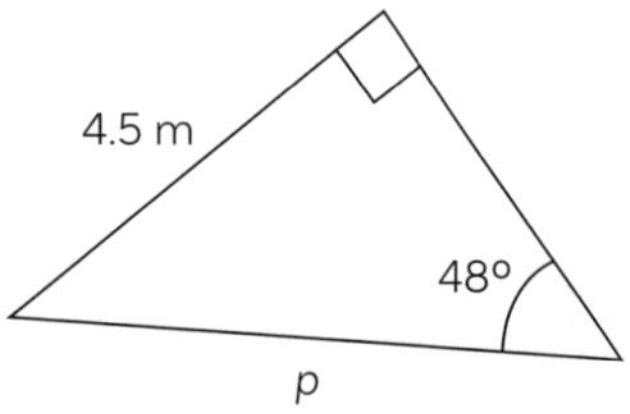

4.5 m

48°

p

6

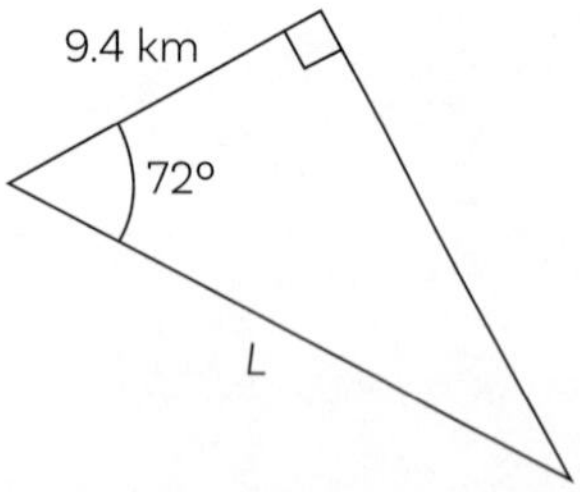

9.4 km

72°

L

7

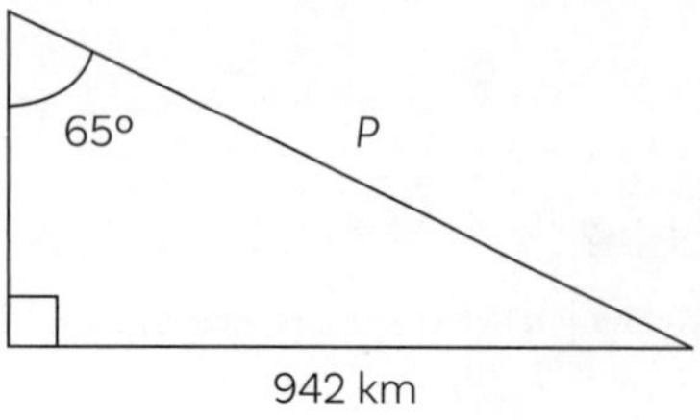

65°

P

942 km

 ISBN: 9780170484084

Finding sides using tangents

This involves using either of the algebraic methods used earlier.

Example 1: Calculate the length marked x.

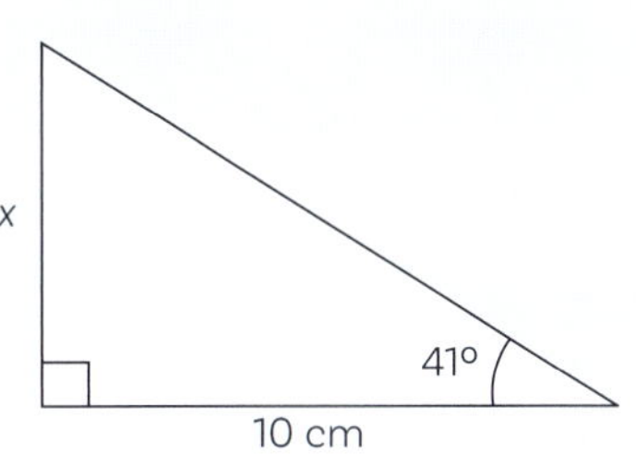

Step 1: **Label** the sides that are involved in the problem with 'A', 'O' and 'H'.

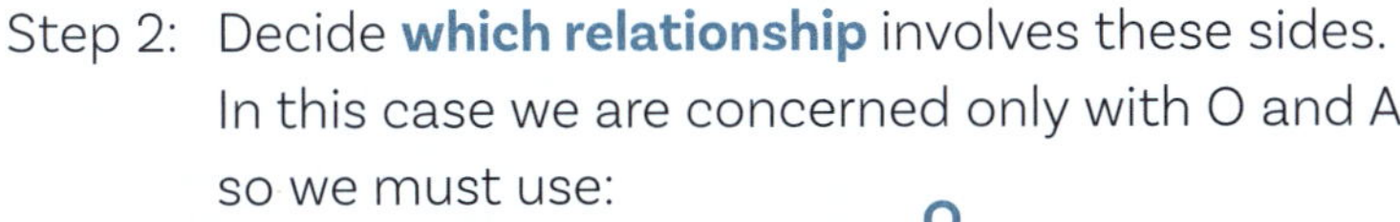

Step 2: Decide **which relationship** involves these sides. In this case we are concerned only with O and A so we must use:

$$\tan \theta = \frac{\mathbf{O}}{\mathbf{A}}$$

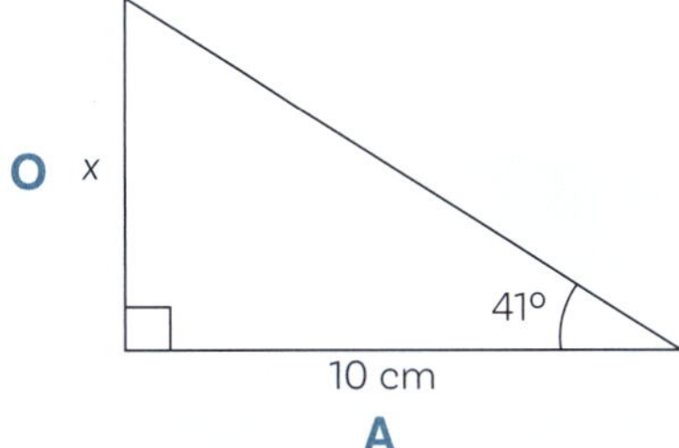

Step 3: **Substitute** the values from the triangle.

$$\tan 41^\circ = \frac{x}{10}$$

$$x = 10 \tan 41^\circ$$

$$x = 8.693 \text{ cm}$$

Step 4: **Think about your answer — does it seem reasonable?** In this case, x must be less than 10 cm because x is opposite the smallest angle (41°), so an answer of 8.693 cm is reasonable.

Example 2: Calculate the length marked x.

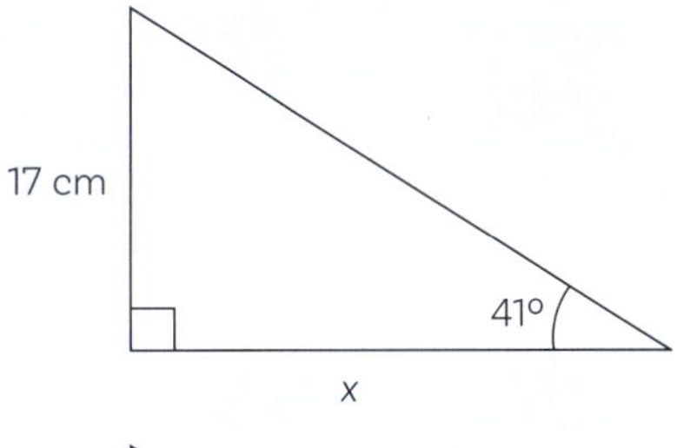

Step 1: **Label** the sides that are involved in the problem with 'A', 'O' and 'H'.

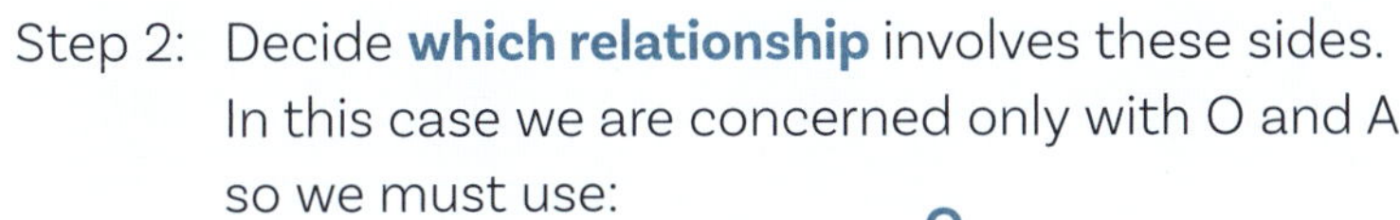

Step 2: Decide **which relationship** involves these sides. In this case we are concerned only with O and A so we must use:

$$\tan \theta = \frac{\mathbf{O}}{\mathbf{A}}$$

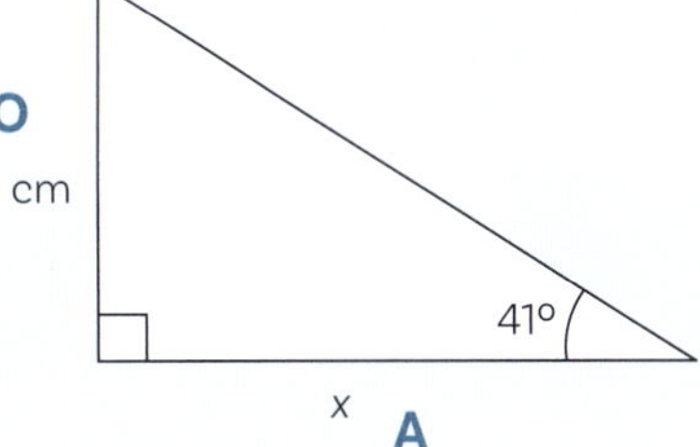

Step 3: **Substitute** the values from the triangle.

$$\tan 41^\circ = \frac{17}{x}$$

$$x \tan 41^\circ = 17$$

$$x = \frac{17}{\tan 41^\circ}$$

$$x = 19.56 \text{ cm}$$

Step 4: **Think about your answer — does it seem reasonable?** In this case, x must be more than 17 cm because the 17 cm is opposite the smallest angle (41°), so an answer of 19.56 cm is reasonable.

Calculate the length of the unknown side of each triangle.

1

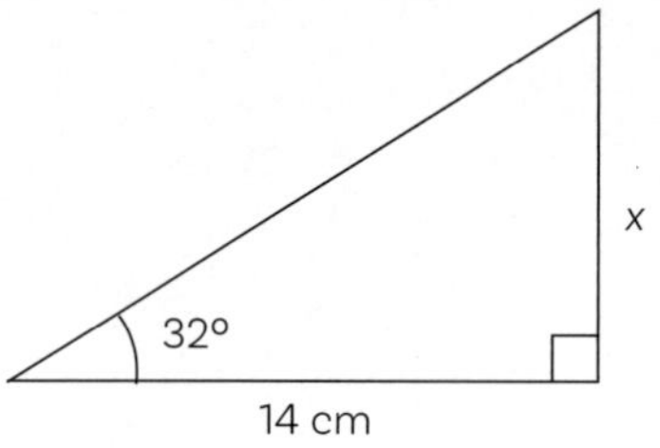

32°
x
14 cm

2

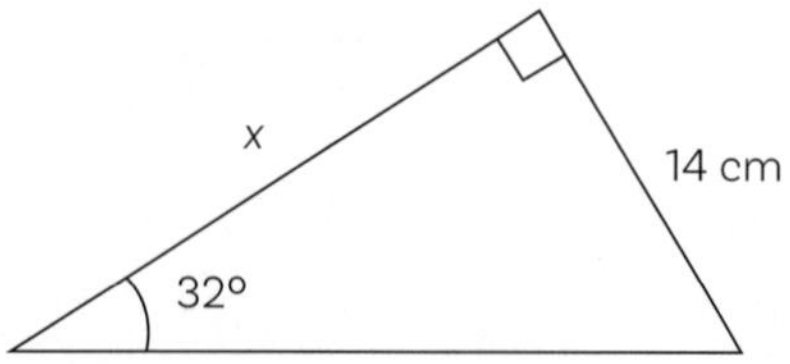

x
14 cm
32°

3

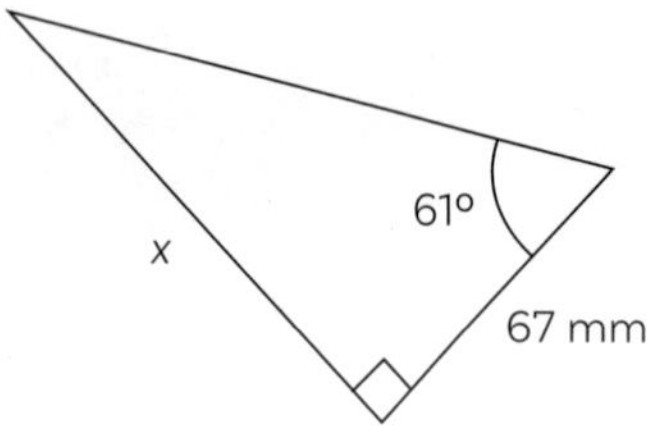

61°
x
67 mm

4

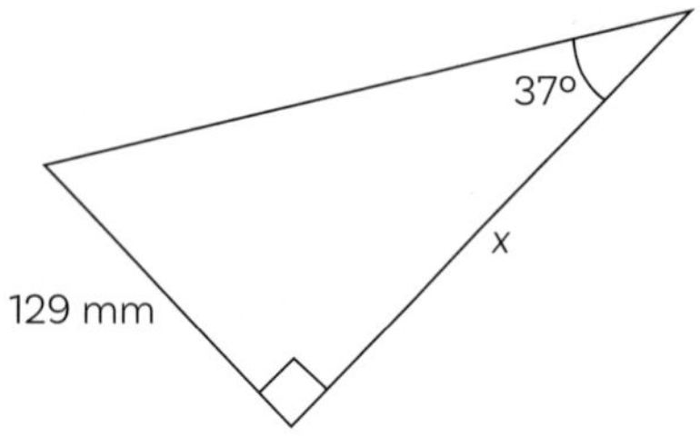

37°
x
129 mm

5

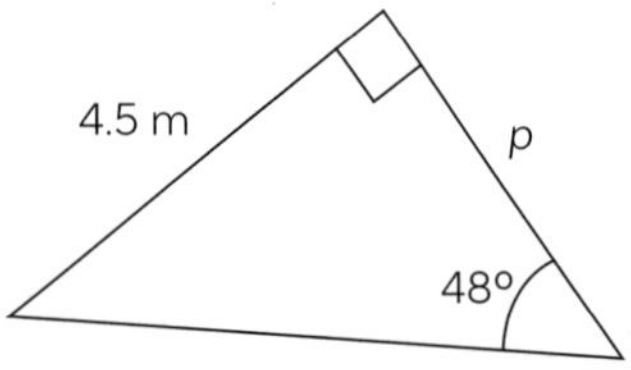

4.5 m
p
48°

6

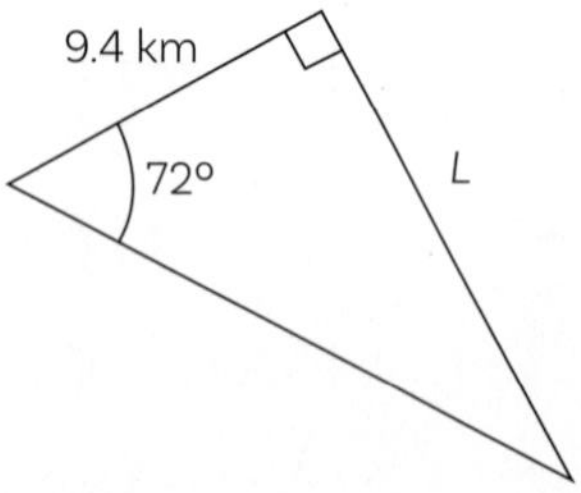

9.4 km
72°
L

7

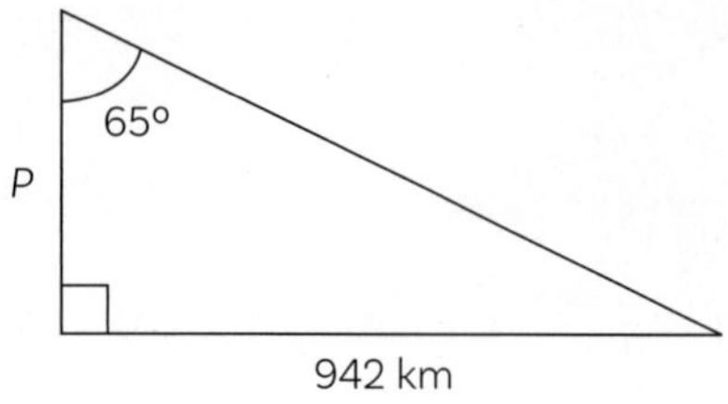

65°
P
942 km

 ISBN: 9780170484084

Finding angles

Once again we use the three rules:

$$\sin\theta = \frac{O}{H} \qquad \cos\theta = \frac{A}{H} \qquad \tan\theta = \frac{O}{A}$$

Example: Calculate the angle marked θ.

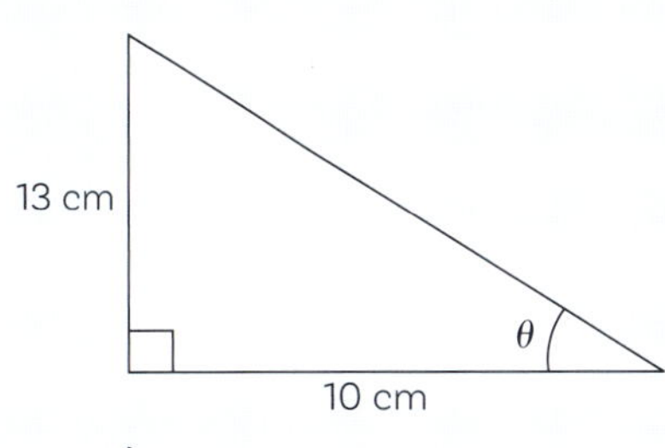

Step 1: **Label** the sides that are involved in the problem with 'A', 'O' and 'H'.

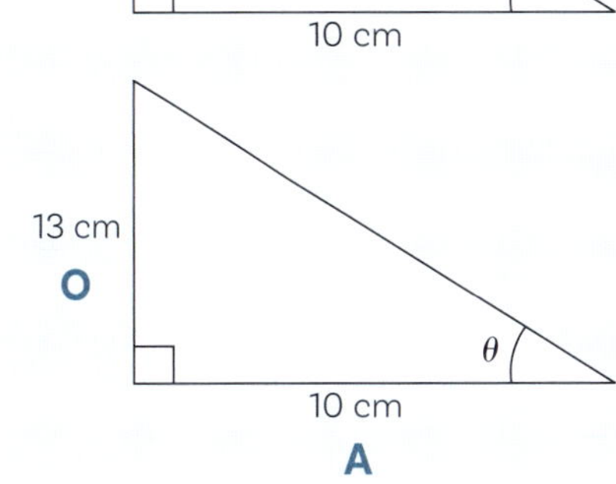

Step 2: Decide **which relationship** involves these sides. In this case we are concerned only with O and A so we must use:

$$\tan\theta = \frac{O}{A}$$

Step 3: **Substitute** the values from the triangle.

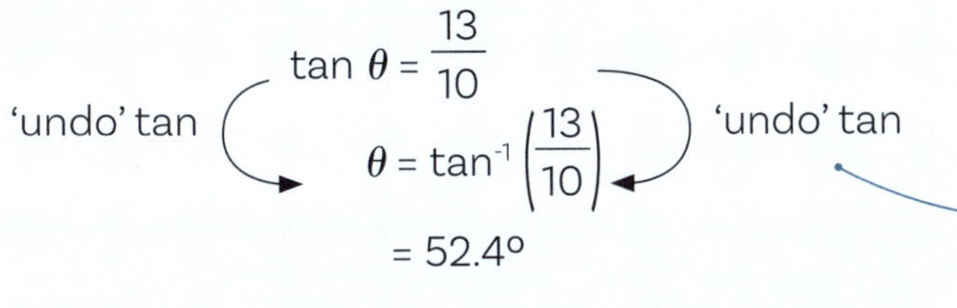

$$\tan\theta = \frac{13}{10}$$

'undo' tan

$$\theta = \tan^{-1}\left(\frac{13}{10}\right)$$

'undo' tan

$$= 52.4^{\circ}$$

Use the 'inverse tan' button on your calculator. Don't forget the **brackets**.

Step 4: **Think about your answer — does it seem reasonable?**
In this case, θ must be more than 45° because 13 cm is longer than 10 cm, so an answer of 52.4° is reasonable.

Calculate the angle marked θ in the following triangles.

1

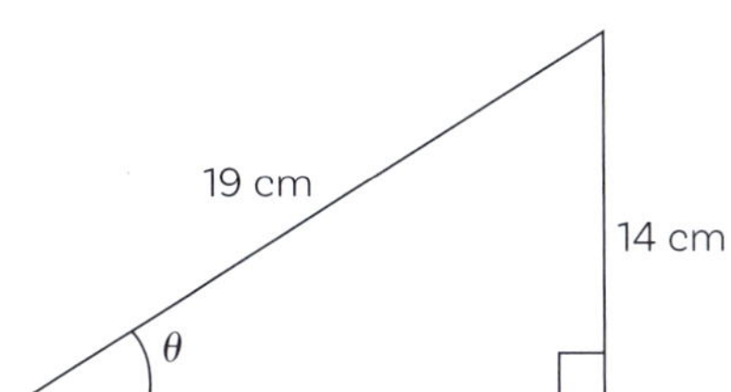

____

____

____

2

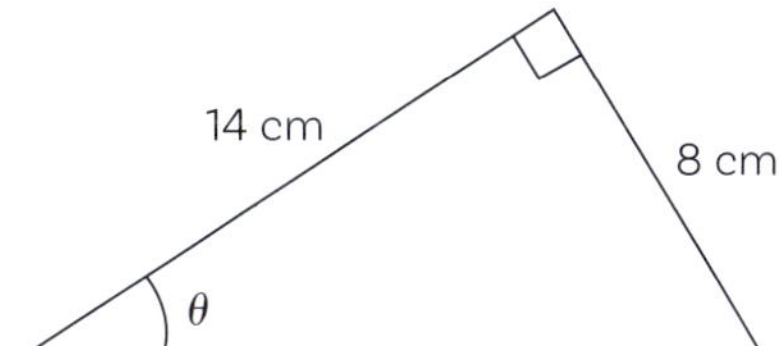

____

____

____

3

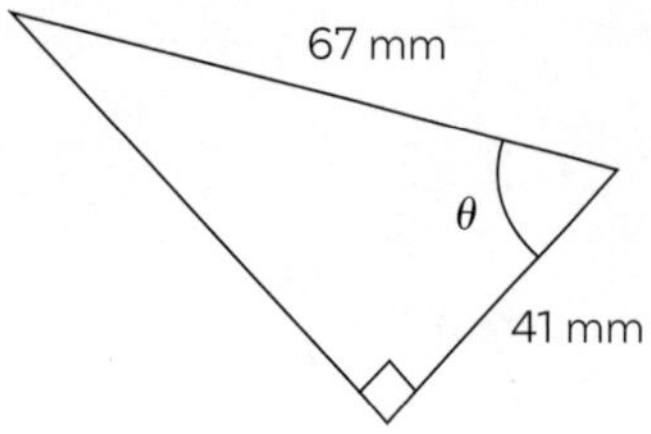

67 mm

θ

41 mm

4

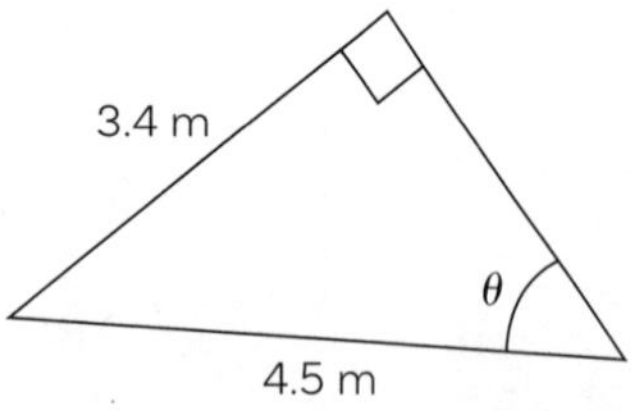

3.4 m

θ

4.5 m

5

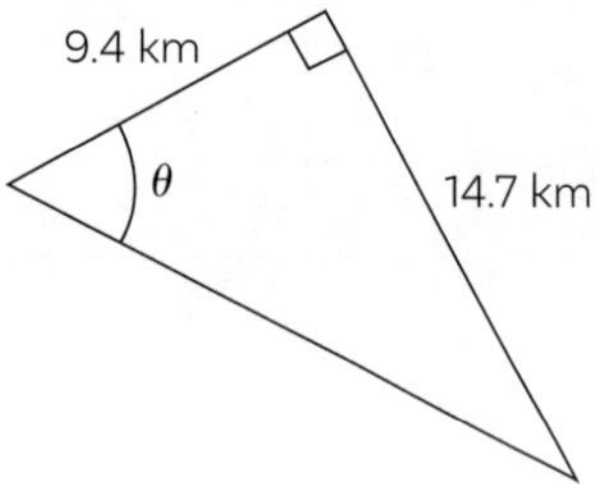

9.4 km

θ

14.7 km

6

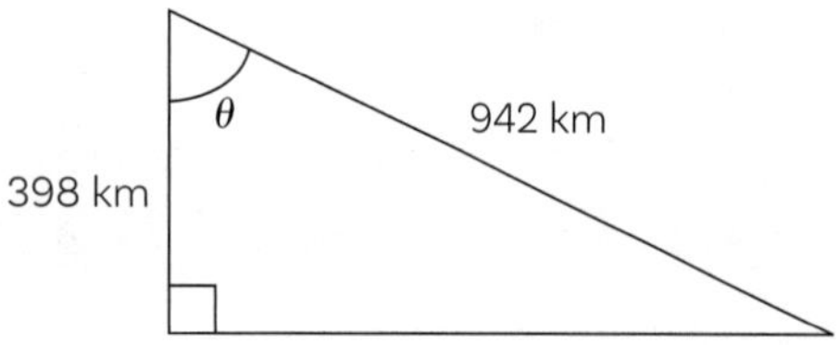

θ

942 km

398 km

7

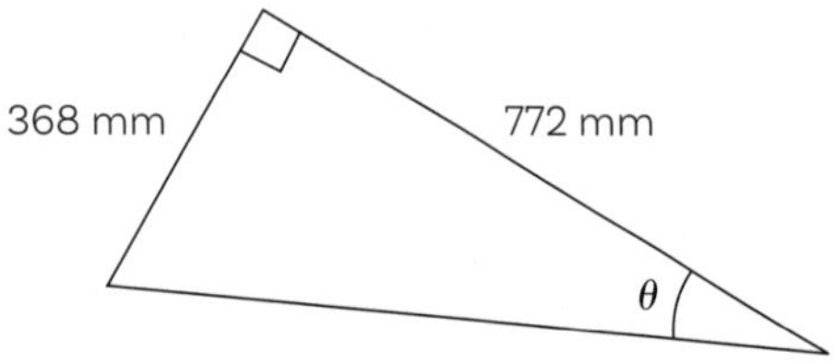

368 mm

772 mm

θ

8

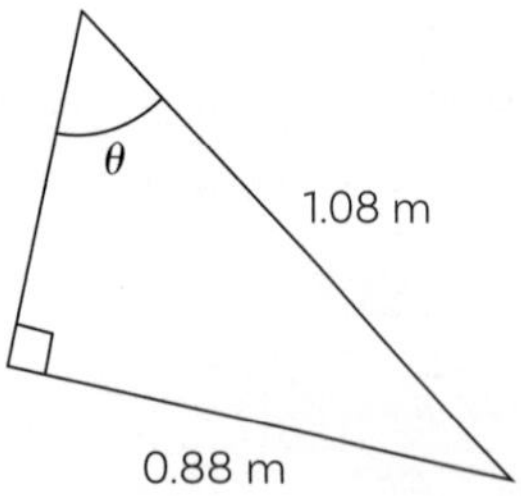

θ

1.08 m

0.88 m

9

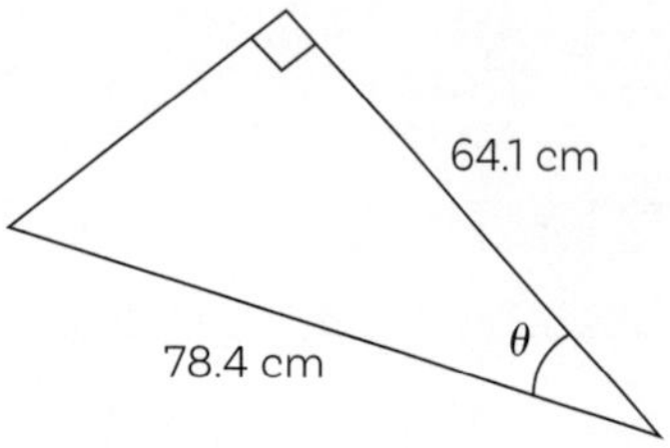

64.1 cm

θ

78.4 cm

 ISBN: 9780170484084

Putting it together — with Pythagoras

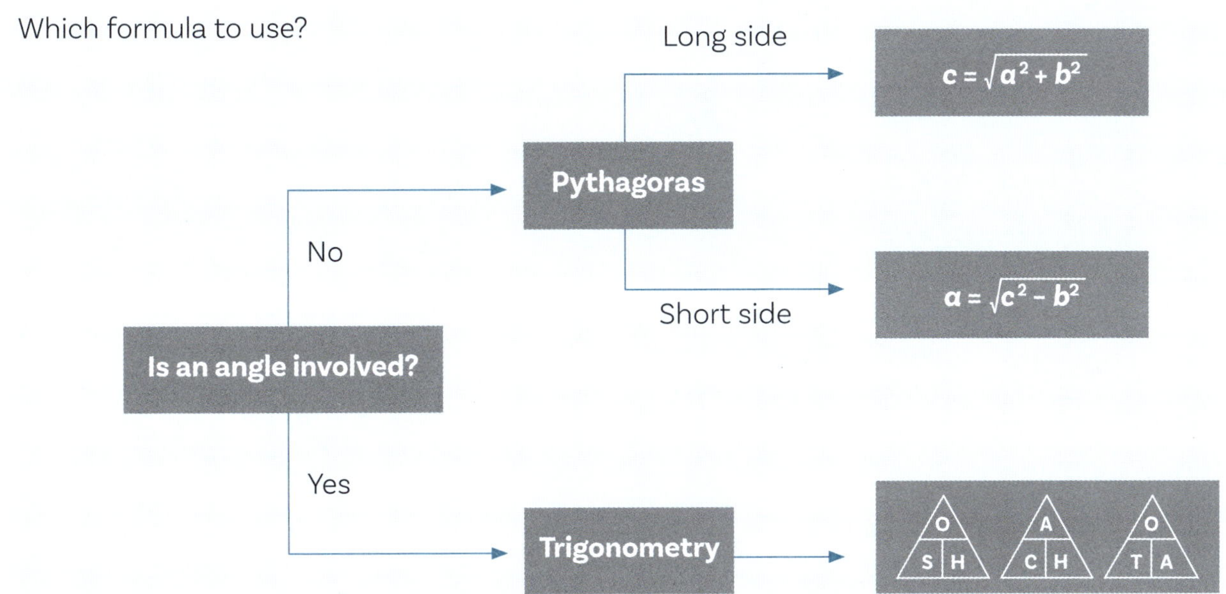

Calculate the unknown sides and/or angles. Round sides to 3 sf and angles to 1 dp.

1

2

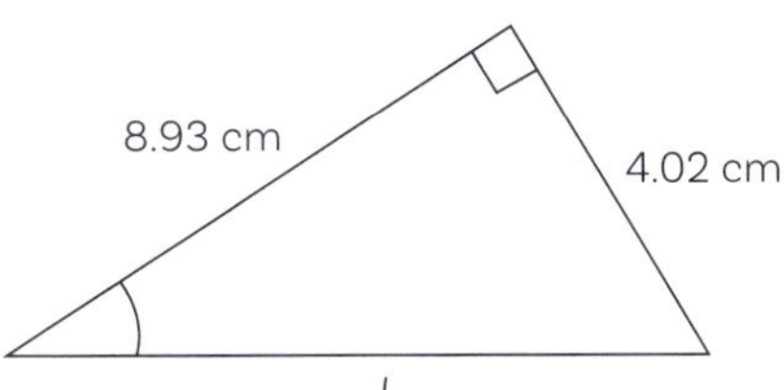

3

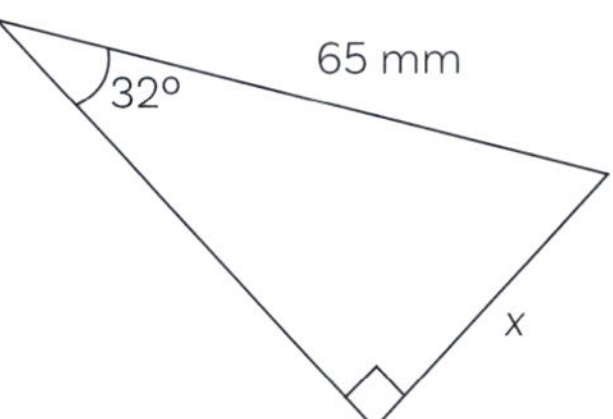

4

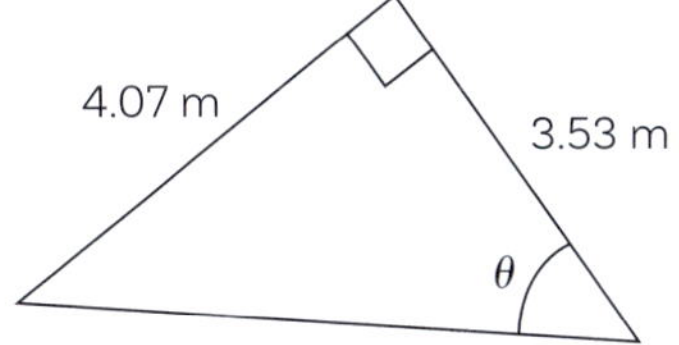

ISBN: 9780170484084

5

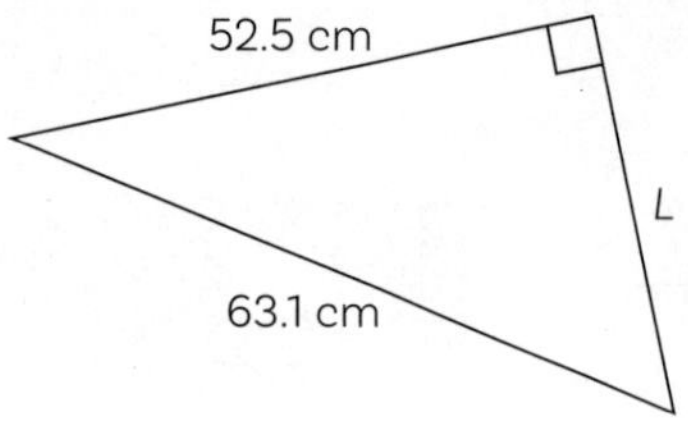

52.5 cm

L

63.1 cm

6

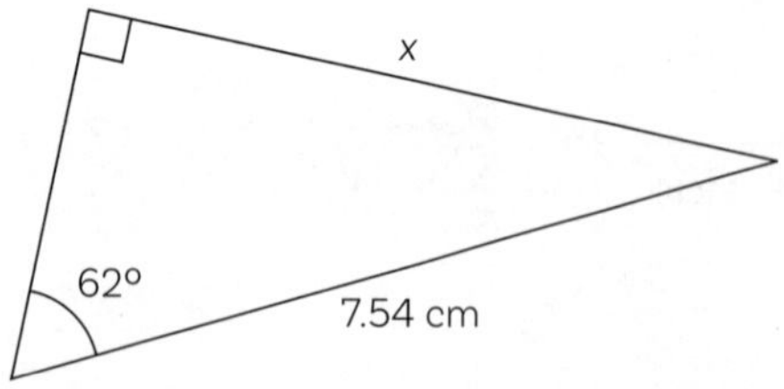

x

62°

7.54 cm

7

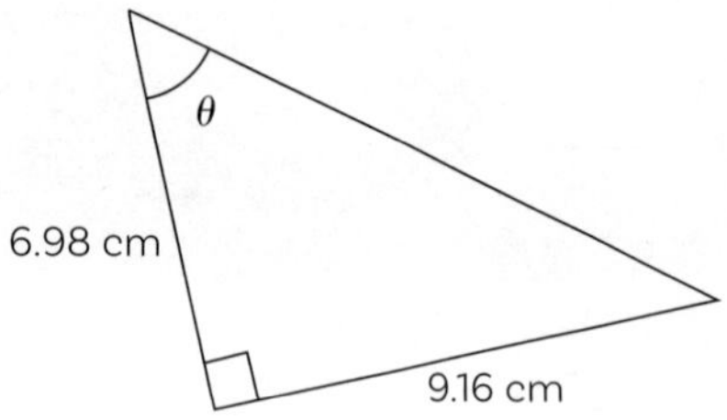

θ

6.98 cm

9.16 cm

8

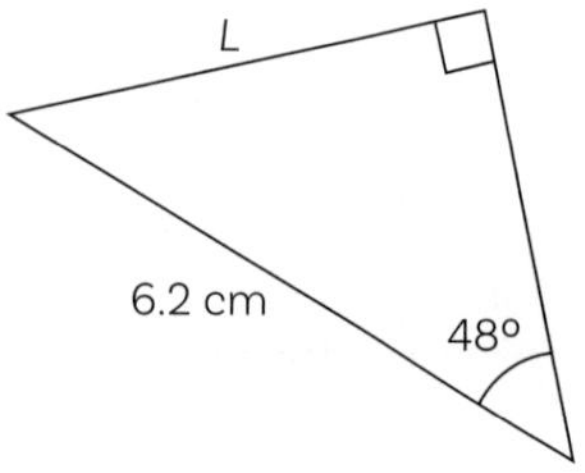

L

6.2 cm

48°

9

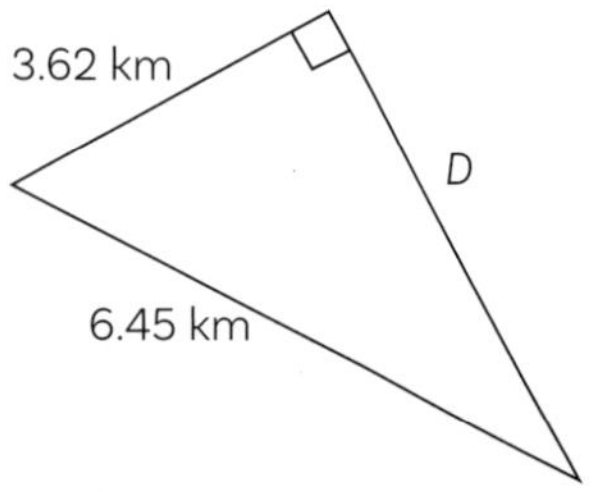

3.62 km

D

6.45 km

10

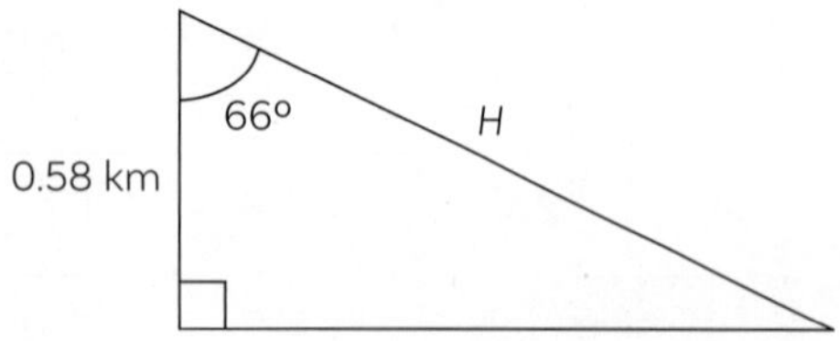

66°

H

0.58 km

11

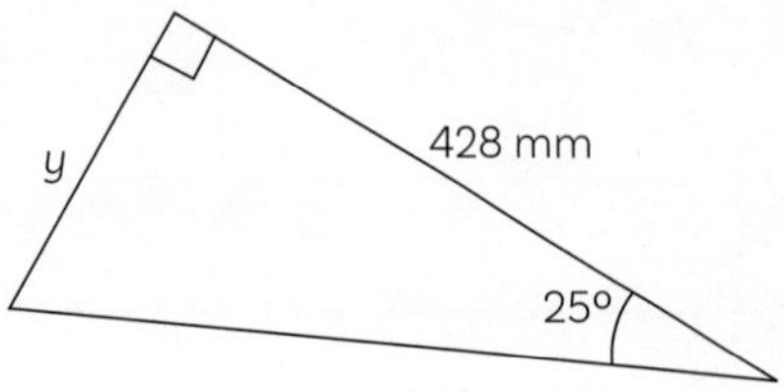

y

428 mm

25°

 ISBN: 9780170484084

3D shapes

- When asked to find the length of a side or an angle in a 3D shape, it is a good idea to **draw the 2D figure** within which it lies.
- These usually involve right-angled triangles, so you can expect to use the Theorem of Pythagoras and trigonometry.

Example: This figure represents a cuboid.

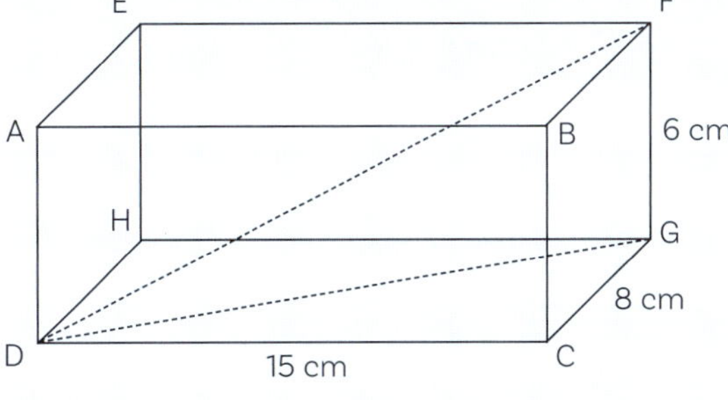

1 Calculate the length of DG.
DG lies in a triangle on the 'floor' of the cuboid.

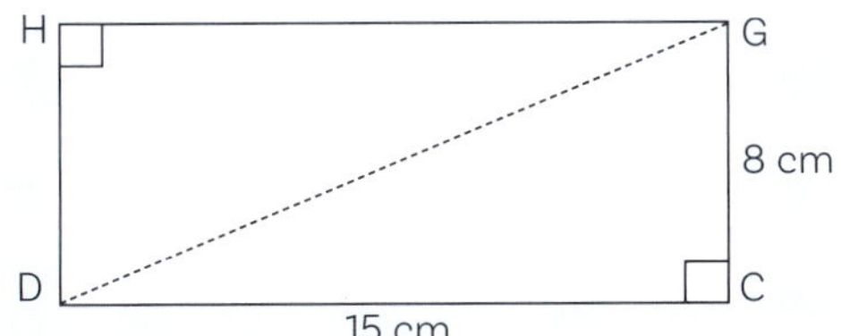

$DG^2 = 15^2 + 8^2$ (Pythagoras)
$\therefore DG = 17$ cm

2 Calculate the length of DF.
DF lies on the triangle DFG.

$DF^2 = 17^2 + 6^2$ (Pythagoras)
$\therefore DF = 18.03$ cm

3 Calculate the angle between the line DF and the plane DHGC.
∠FDG lies on triangle DFG.

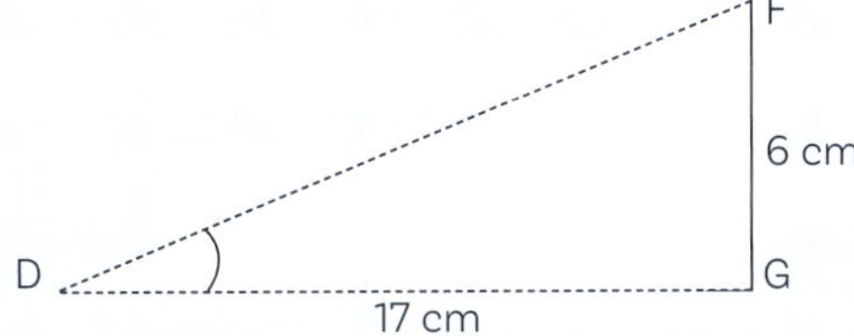

$\tan FDG = \frac{6}{17}$

$\therefore \angle FDG = 19.4°$

4 Calculate the angle between the plane HFD and the plane DHGC.

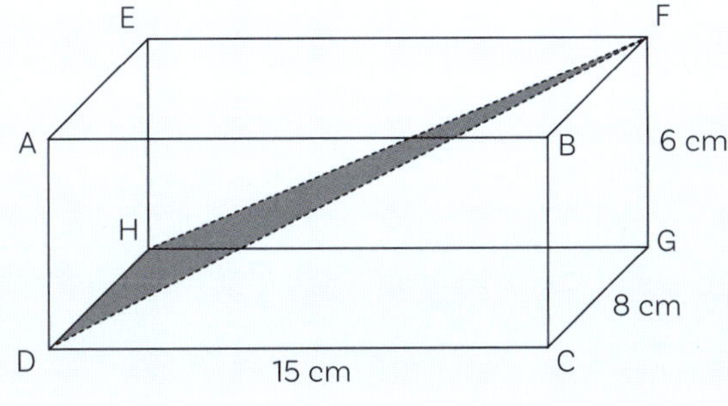

∠FHG lies in the triangle FGH.

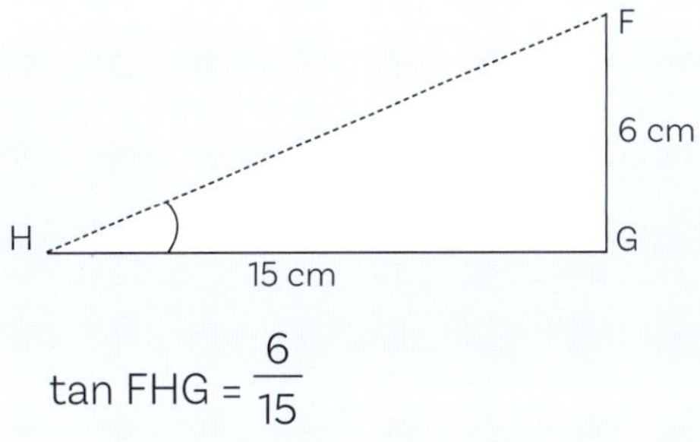

$\tan FHG = \frac{6}{15}$

$\therefore \angle FHG = 21.8°$

Calculate the missing angles and sides for the following.

1 This figure shows a cuboid (box).

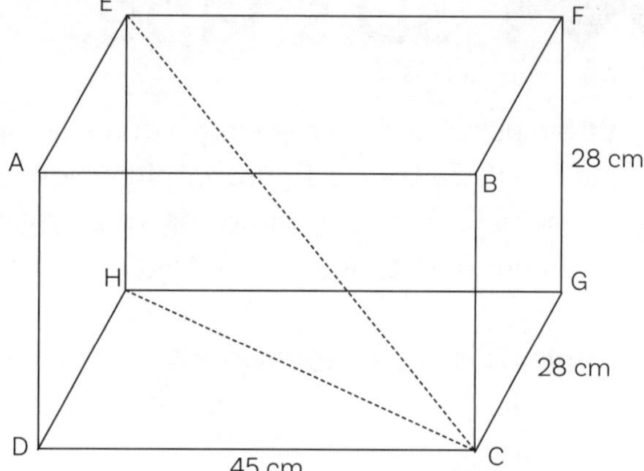

a Calculate the length of HC.

b Calculate the length of EC.

c Calculate the angle between the line EC and the plane HGCD.

d Calculate the angle between the plane ECH and plane EFGH.

e Calculate the angle between the plane EHC and plane AEHD.

 ISBN: 9780170484084

Bearings

- Bearings are used in navigation to define **direction** in a **horizontal** plane.
- Direction can be given in terms of north, south, east and west.

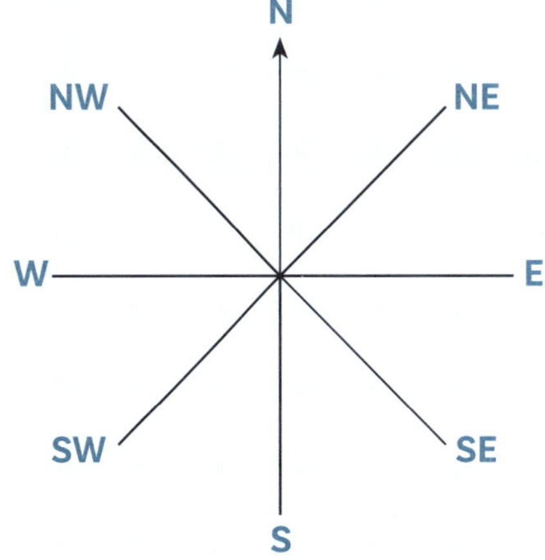

- Bearings are measured in **degrees** in a **clockwise direction from north** (**000° or 360°**).
- All bearings must have **three digits**.

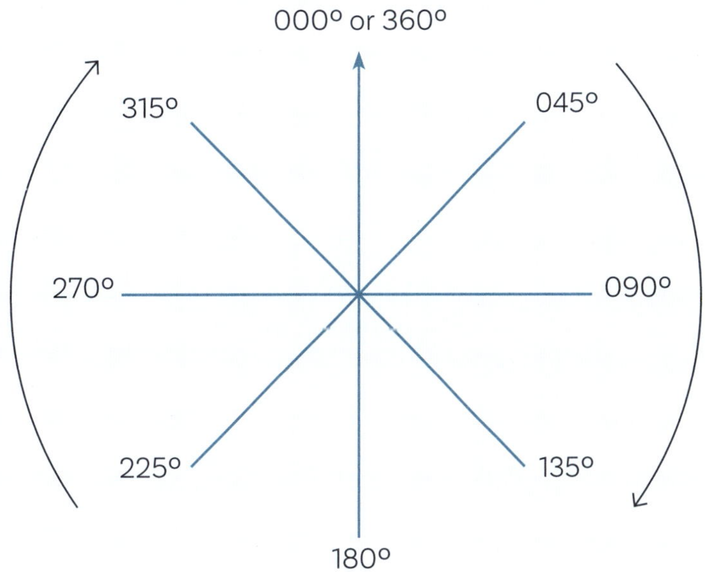

Examples:

1

N

074°

Notice that the line must have direction: in this case an arrow indicates the direction.

2

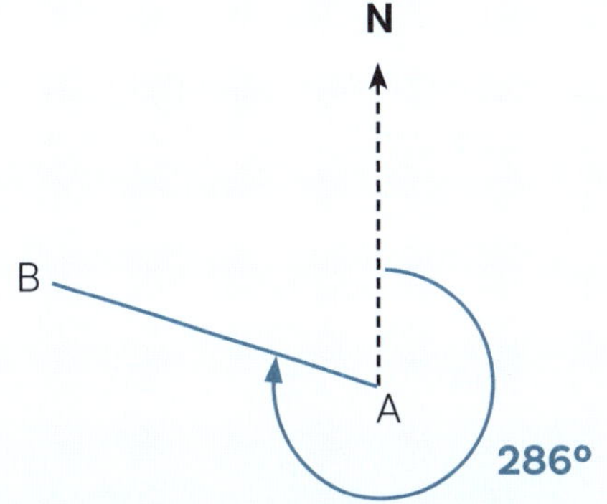

Sometimes the direction will be indicated by words, e.g. 'from A to B'.

ISBN: 9780170484084

Without using a protractor, match the bearings to each of the diagrams. They are either in the direction of the arrow, or from A to B. The dashed line points to the north.

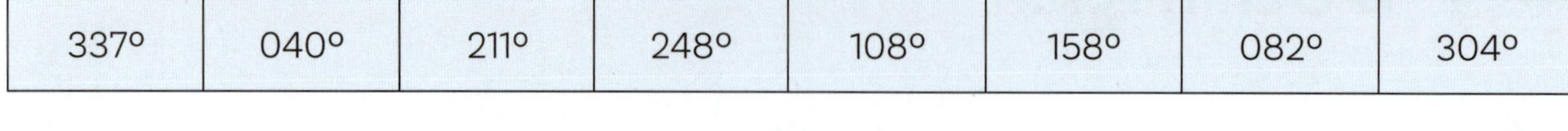

337°	040°	211°	248°	108°	158°	082°	304°

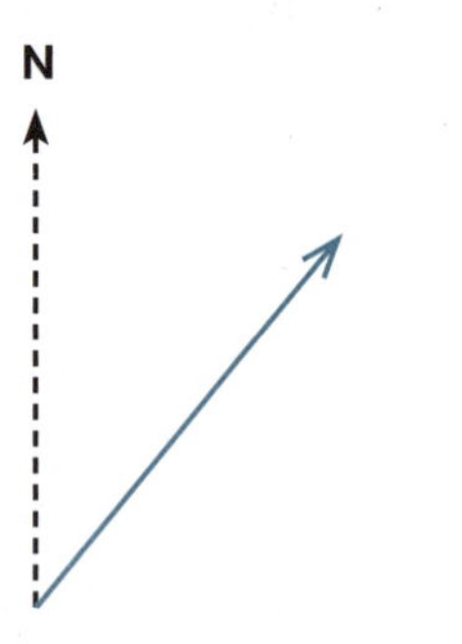

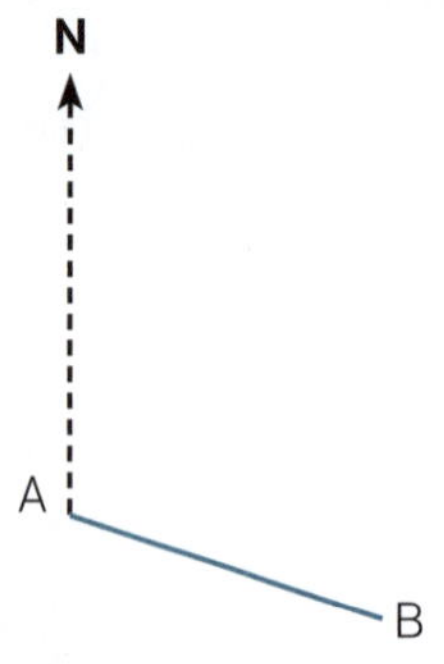

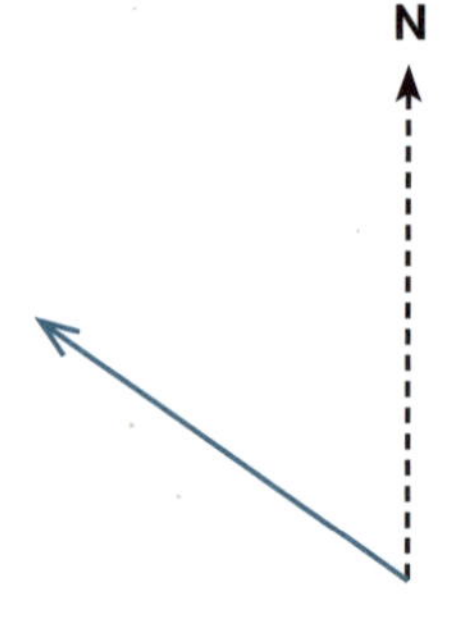

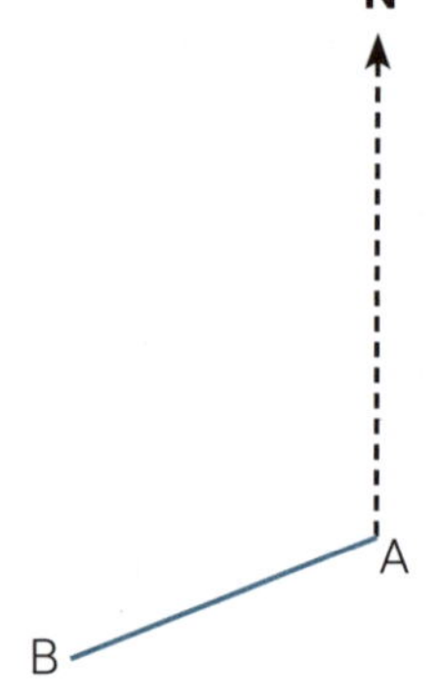

1 ________ **2** ________ **3** ________ **4** ________

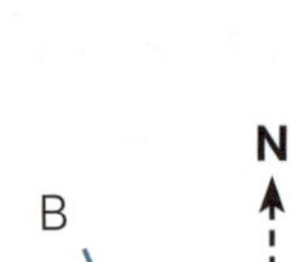

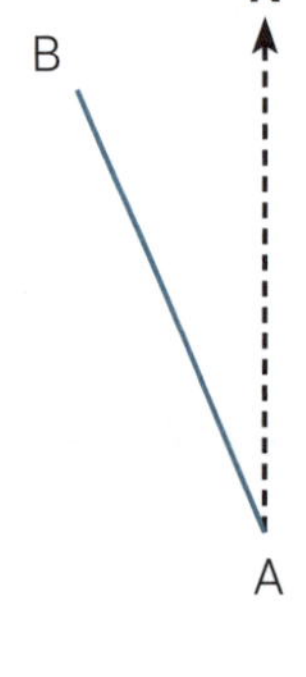

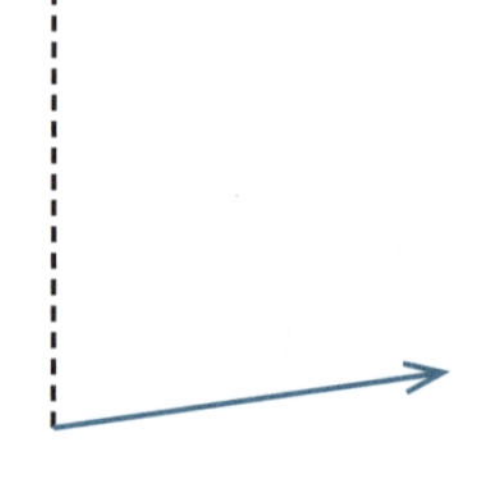

5 ________ **6** ________ **7** ________ **8** ________

9 An orienteering course starts and finishes at point A. Write the bearings for each stage of the course. Select your answers from the list below.

240°	104°
294°	121°
146°	265°
320°	063°
079°	

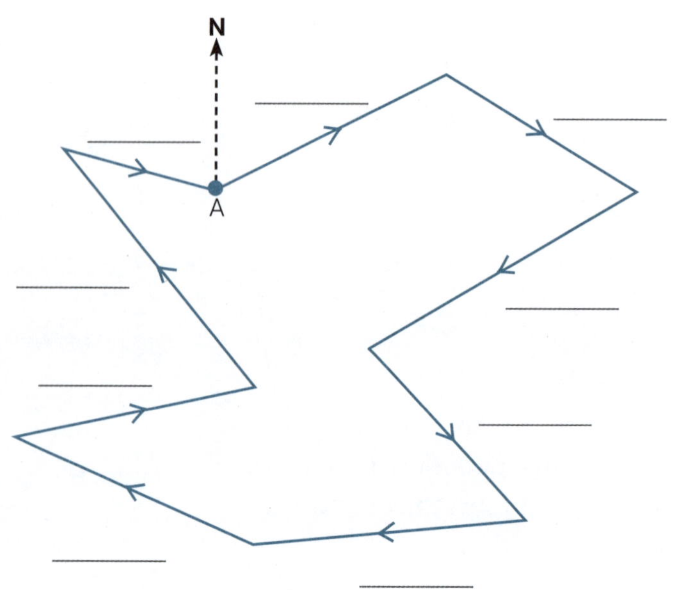

 ISBN: 9780170484084

Unit conversion

Units of length, mass and capacity

Use the following charts to help you convert units.

Length

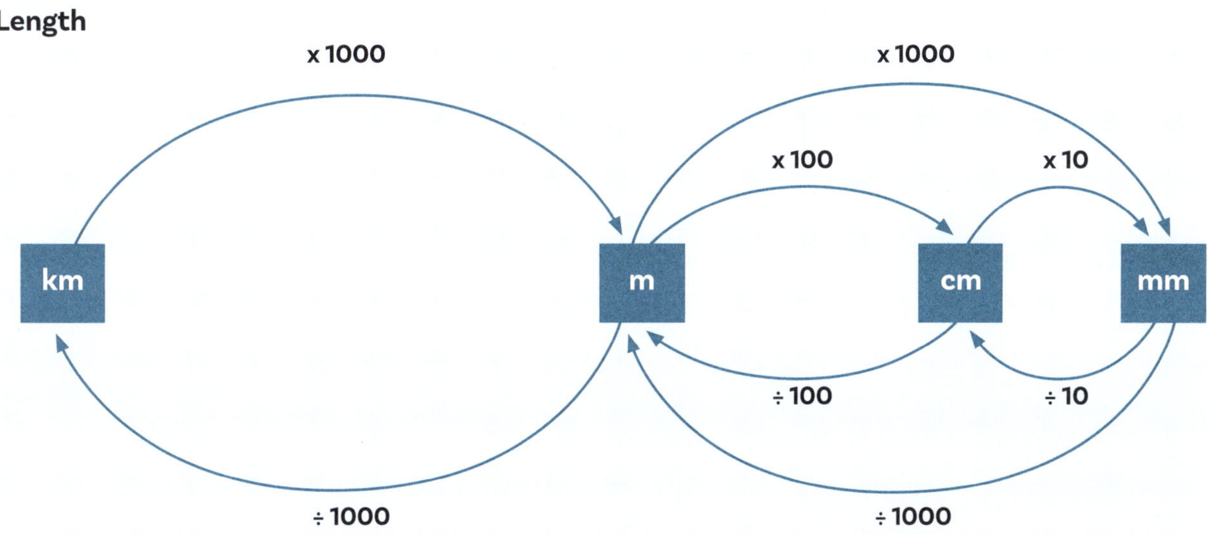

Mass

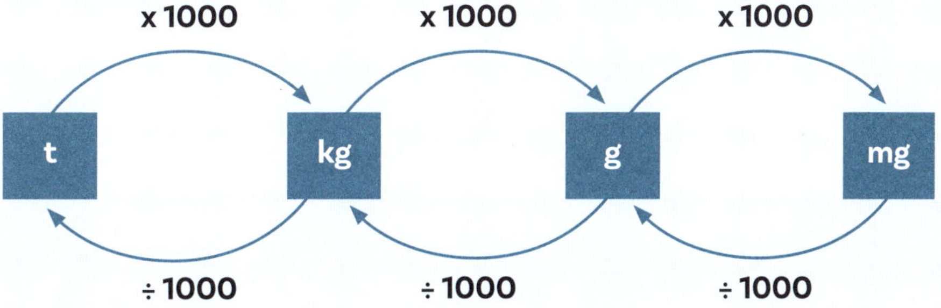

Capacity (fluids)

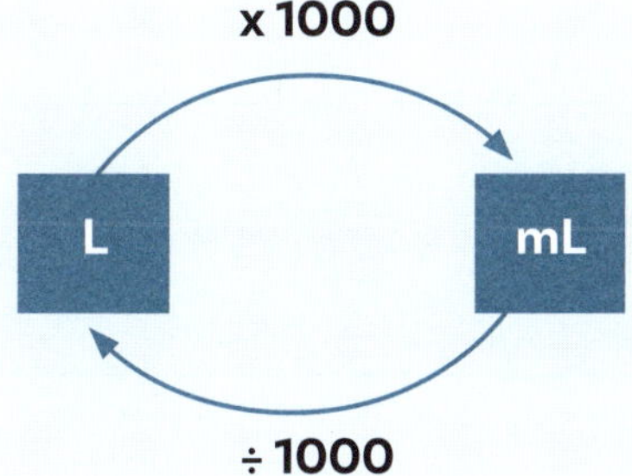

ISBN: 9780170484084

Convert the following.

1 12 cm = ____________ mm

2 3560 g = ____________ kg

3 2950 mL = ____________ L

4 8.9 km = ____________ m

5 3094 kg = ____________ t

6 0.25 L = ____________ mL

7 0.71 g = ____________ mg

8 3 m = ____________ km

9 0.28 t = ____________ kg

10 1860 cm = ____________ km

11 95 mL = ____________ L

12 298 000 mg = ____________ kg

Convert all quantities to the units indicated and then complete the calculations.

13 3.6 m + 82 cm = ____________

____________ m

14 0.86 L + 290 mL = ____________

____________ L

15 850 g + 60 g + 2.9 kg = ____________

____________ kg

16 6300 m + 0.12 km = ____________

____________ km

17 (19 x 38 kg) + 0.6 t = ____________

____________ t

18 5.73 m + 87 mm ____________

____________ m

Choose the most suitable units for the following and then complete the calculations.

19 12.5 km + 23 500 cm + 730 m

____

____

20 2800 mL + 1.25 L – 590 mL

____

____

21 4260 kg – 0.01 t + 1.25 t

____

____

22 29.3 km – 948 m + 78 m

____

____

ISBN: 9780170484084

Perimeter

1 Figures with lines

- The perimeter of a shape is the distance around the outside.
- It is often easier to start at one corner of the shape and add each side until you get back to where you started.
- None of these shapes is drawn to scale.

Example 1:

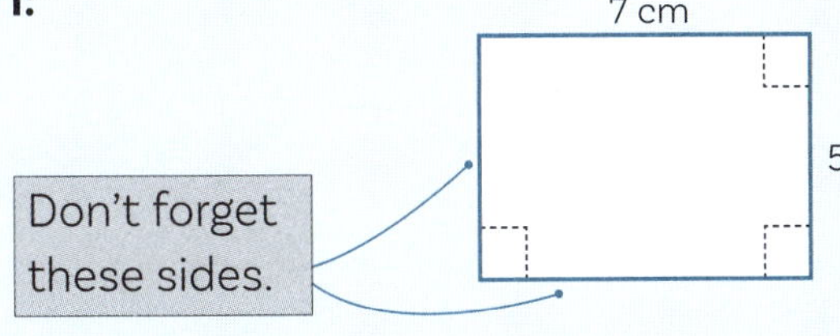

Perimeter = 7 + 5 + 7 + 5
= 24 cm

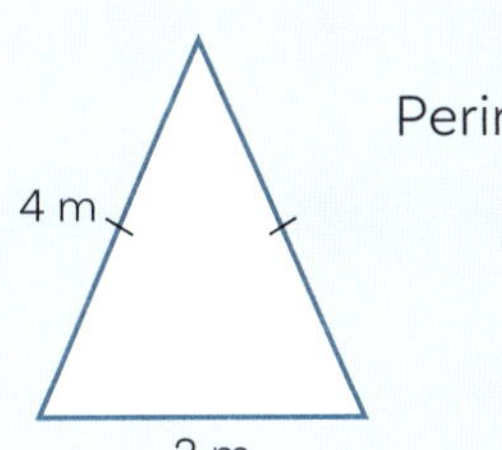

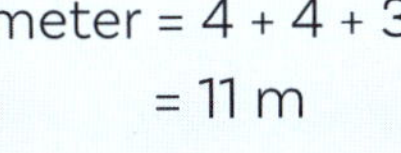

Don't forget these sides.

Opposite sides must be the same, as it is a rectangle.

Example 2:

Perimeter = 4 + 4 + 3
= 11 m

Example 3:

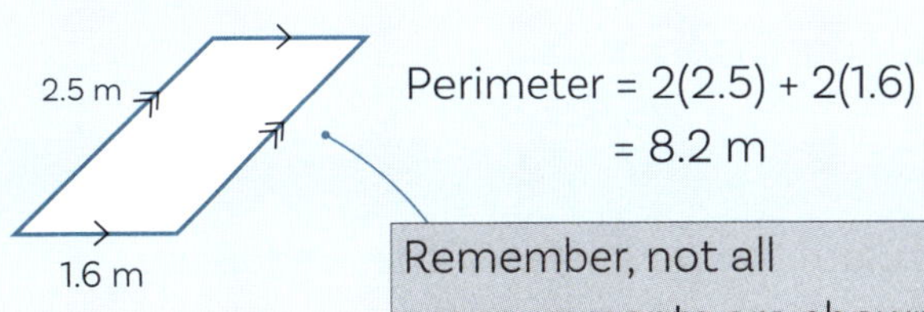

Perimeter = 2(2.5) + 2(1.6)
= 8.2 m

Remember, not all measurements are shown.

Calculate the perimeter of these shapes.

1

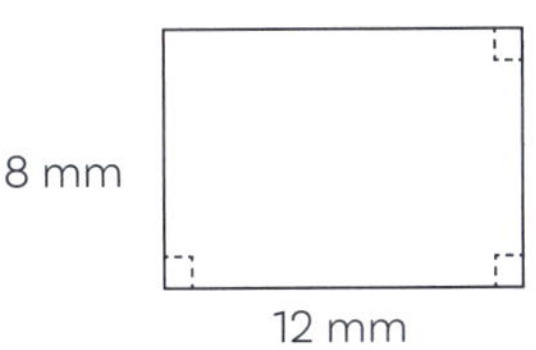

2

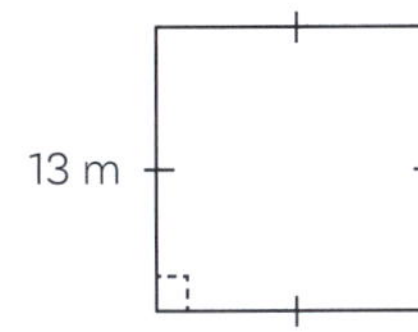

3

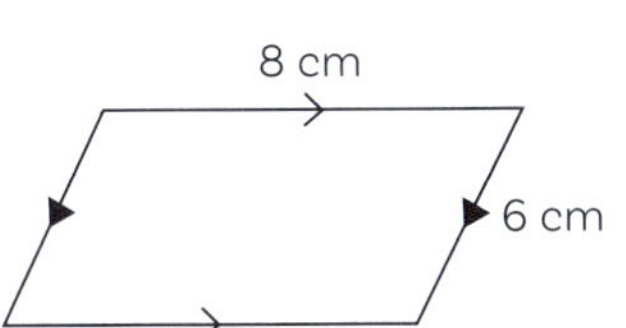

4

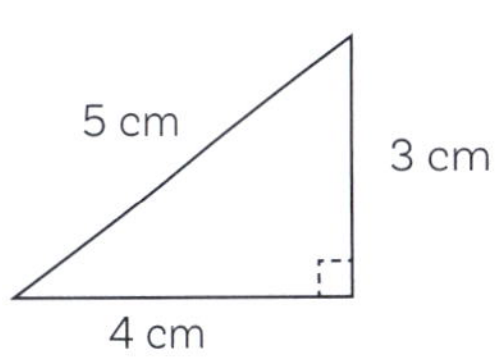

5

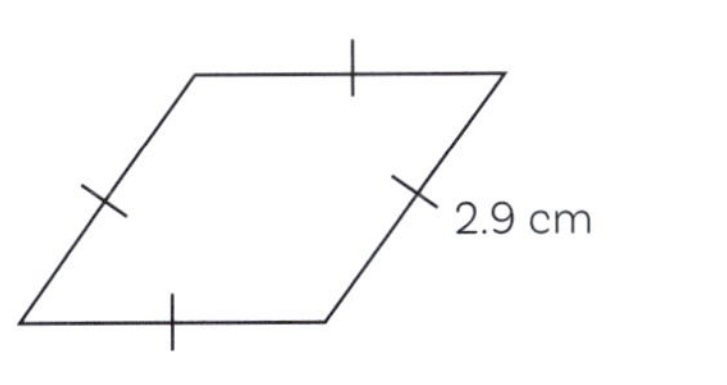

6

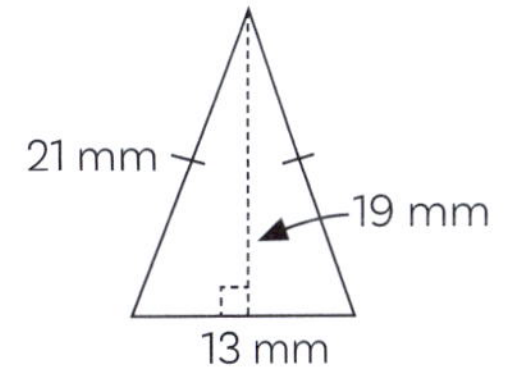

2 Circles – the circumference

The perimeter of a circle is called a circumference.
The formula for this is:

circumference = πd

d stands for **diameter**, which is the distance from one point on a circle, through the centre to another point on the circle.

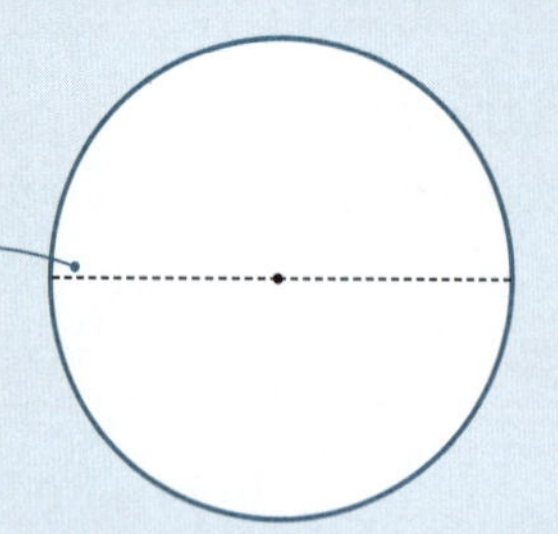

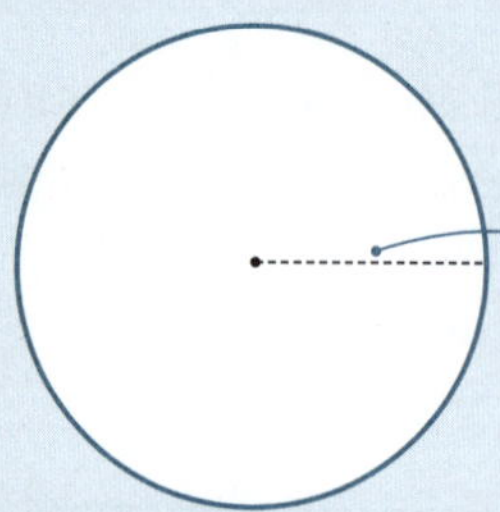

At times you may be given the **radius**. Double it to calculate the diameter.

Example 1:

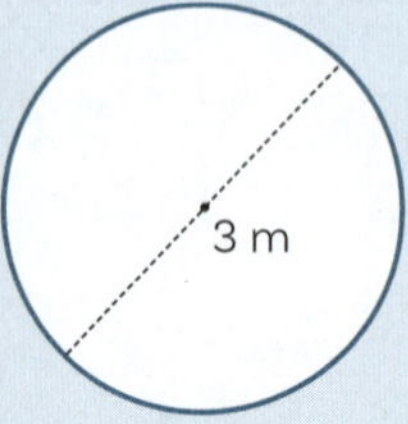

Circumference = $\pi \times 3$
= 9.42 m

Example 2:

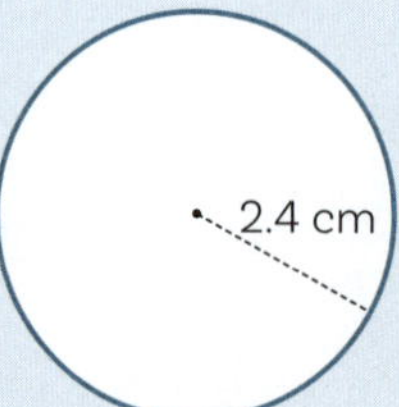

Circumference = $\pi \times (2.4 \times 2)$
= 15.08 cm

Calculate the perimeter of these shapes.

1

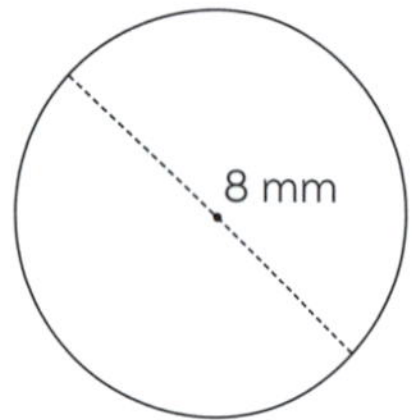

2

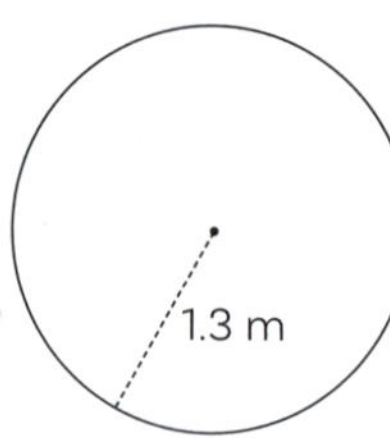

3

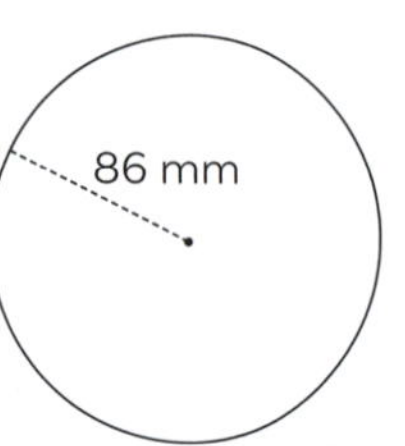

4

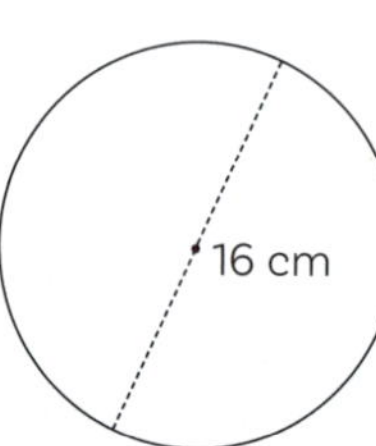

5

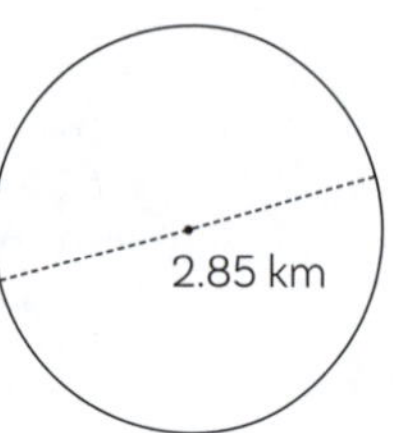

6 A circle has a circumference of 38.64 cm. Calculate its diameter.

ISBN: 9780170484084

3 Compound shapes

- The perimeter of a shape is the distance around the outside.
- Compound shapes are shapes that are made up of other simple shapes.

Example:

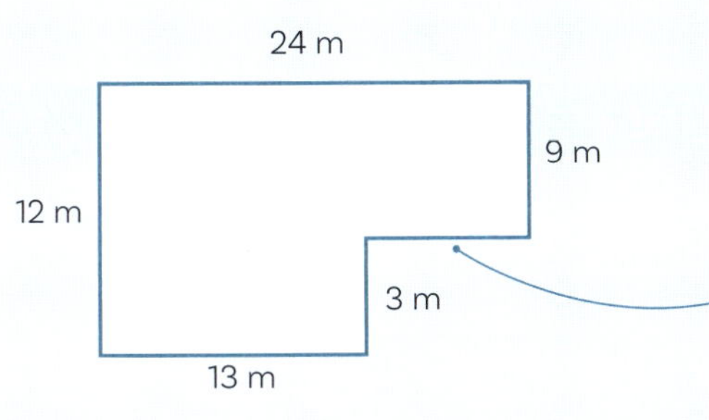

Perimeter = 12 + 24 + 9 + 11 + 3 + 13
= 72 m

This side isn't labelled but you can work out that it is 11 m from the other measurements.

Calculate the perimeter of these shapes.

1

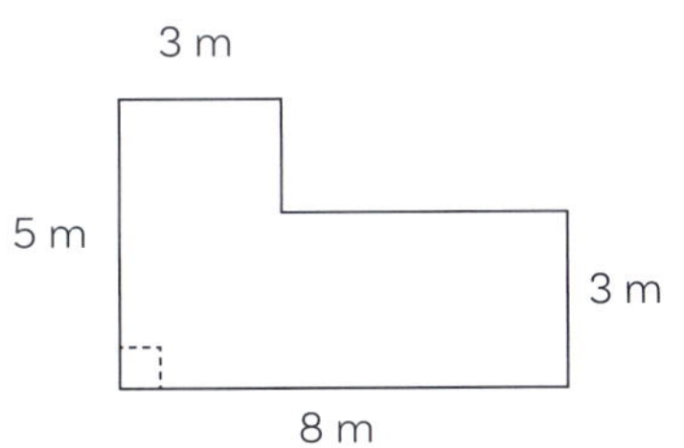

2

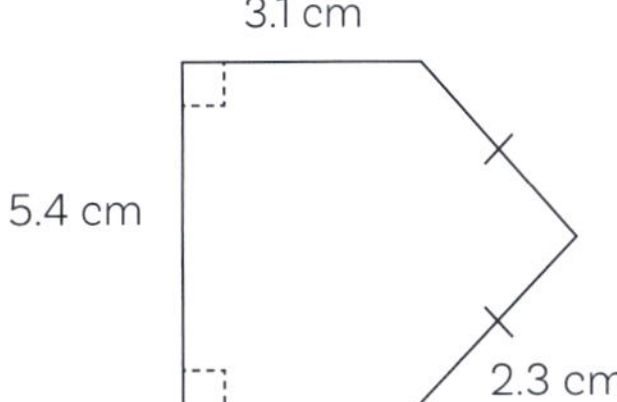

3

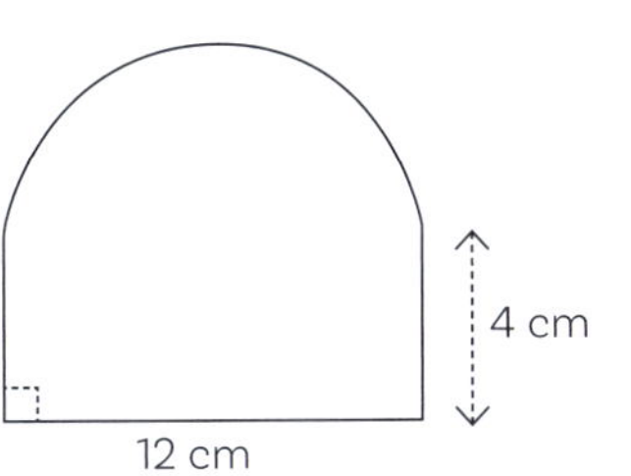

4

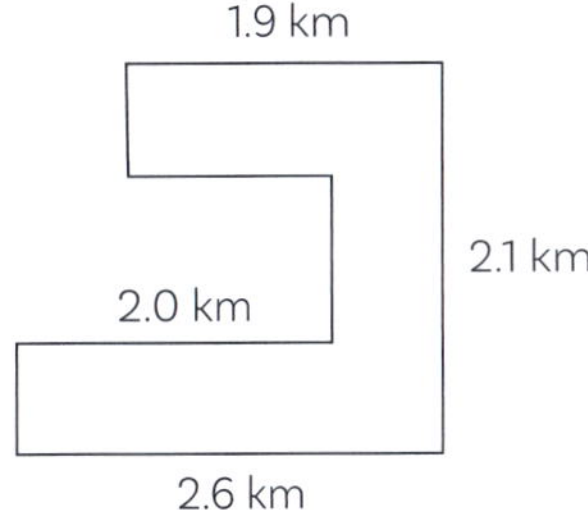

5

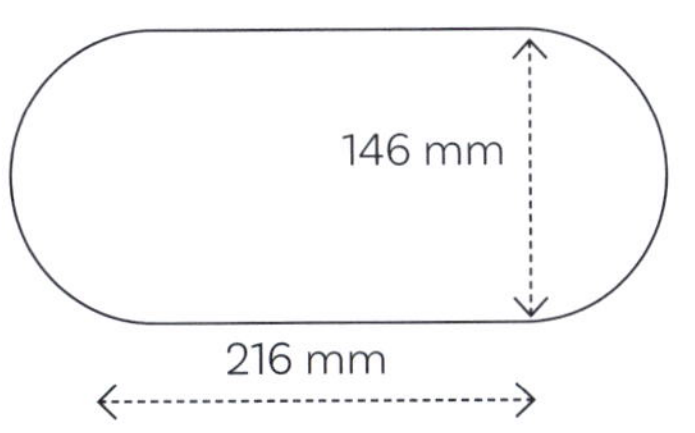

6

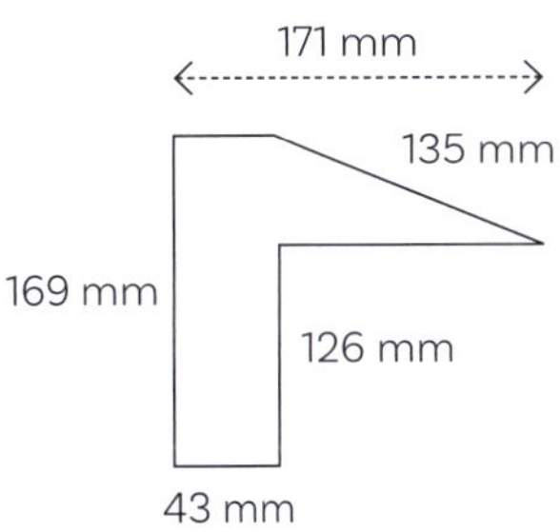

Area

1 Triangles and quadrilaterals

Complete the table.

Shape	Area	Shape	Area
Rectangle h, b	**area = base x height**	Parallelogram and rhombus h, a, b	
Triangle a, h, c, b		Trapezium d, c, h, a, b	

Calculate the area of these shapes.

1

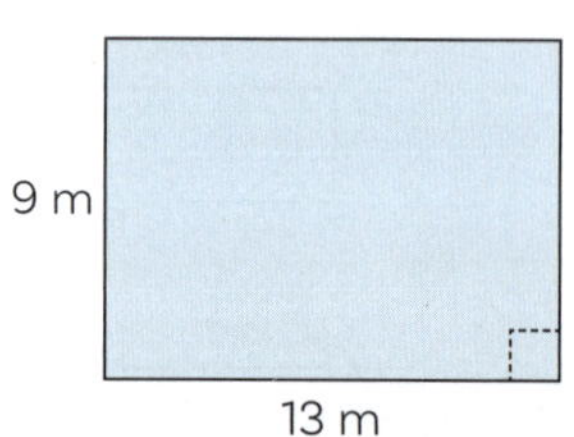

2

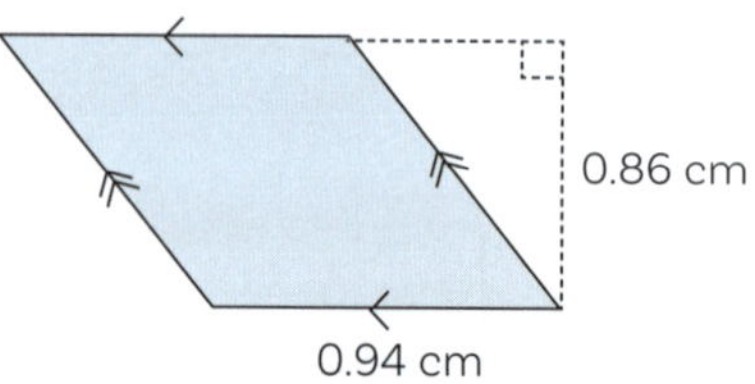

3

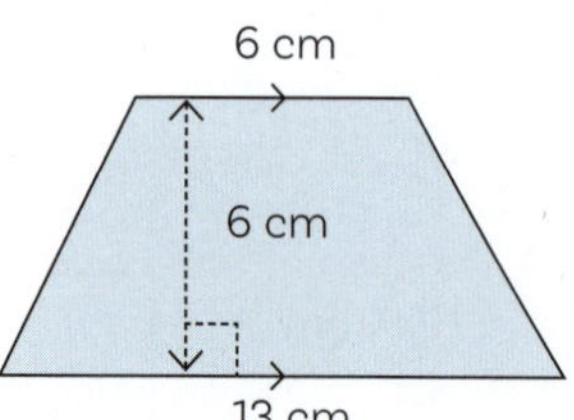

4

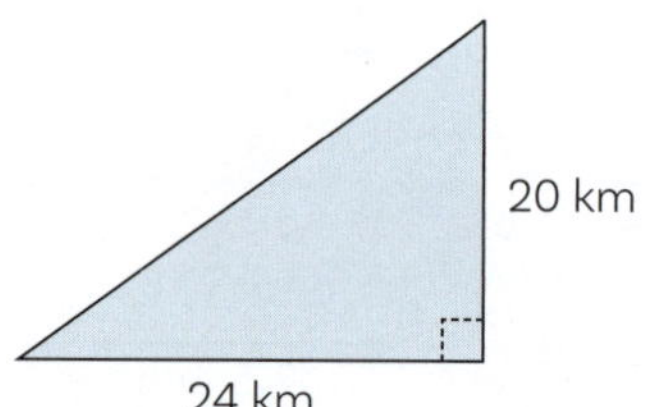

ISBN: 9780170484084

5

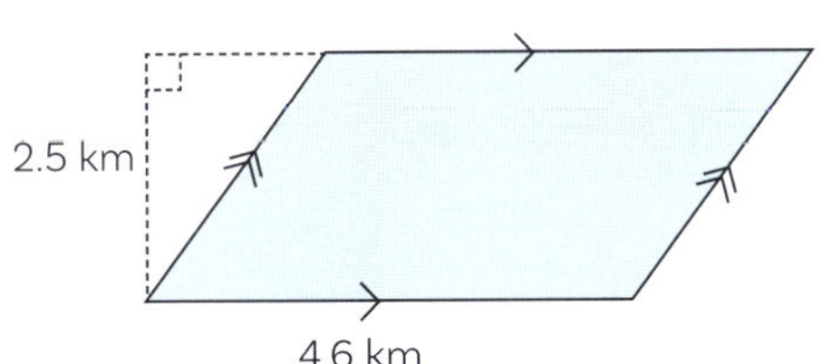

6

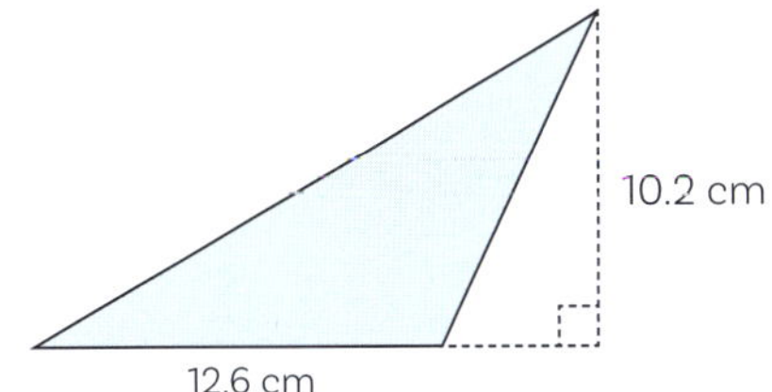

7

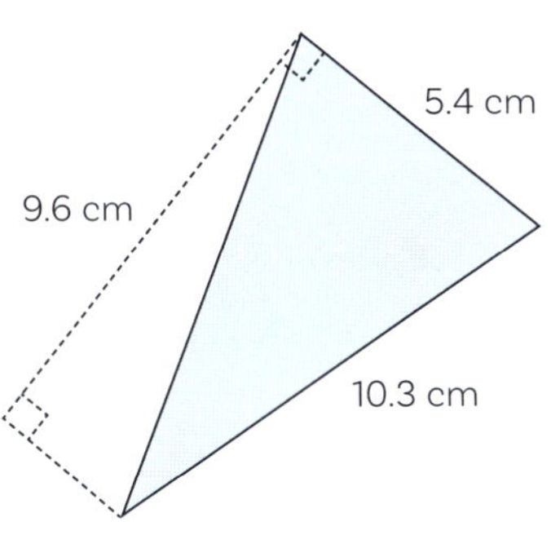

8

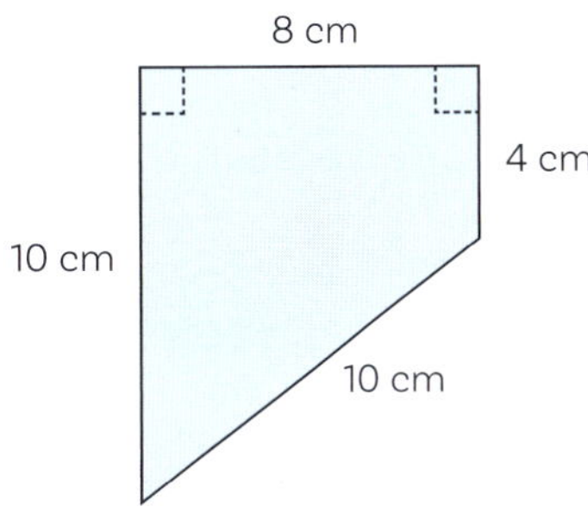

9

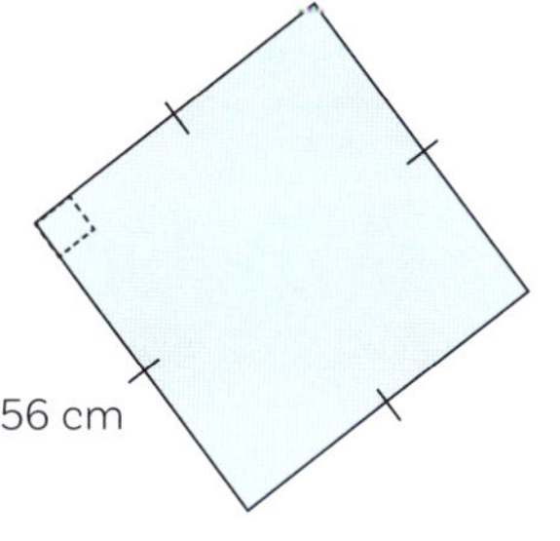

10

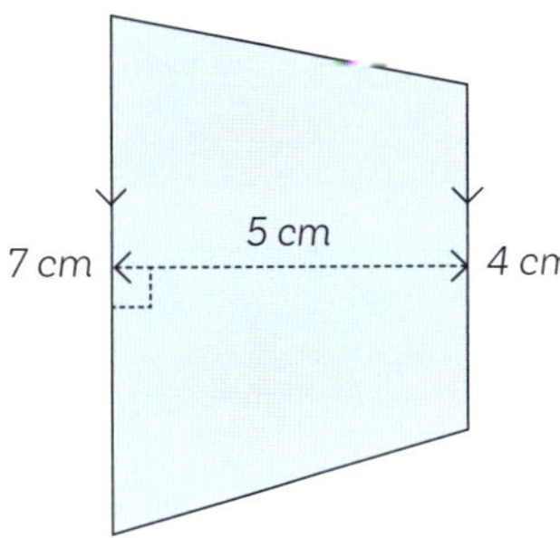

11

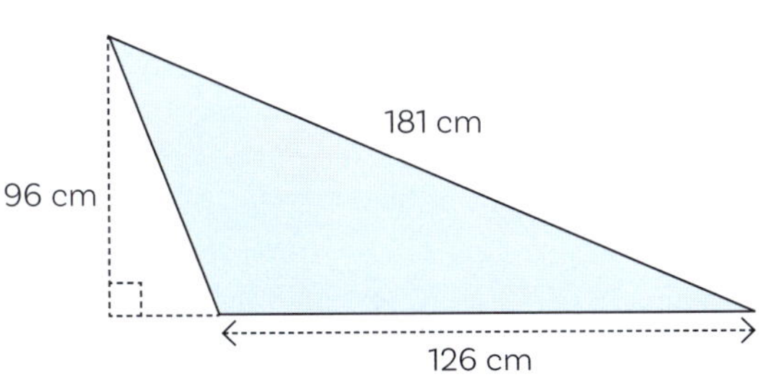

12 The area of this trapezium is 94.5 m^2. Calculated the length x.

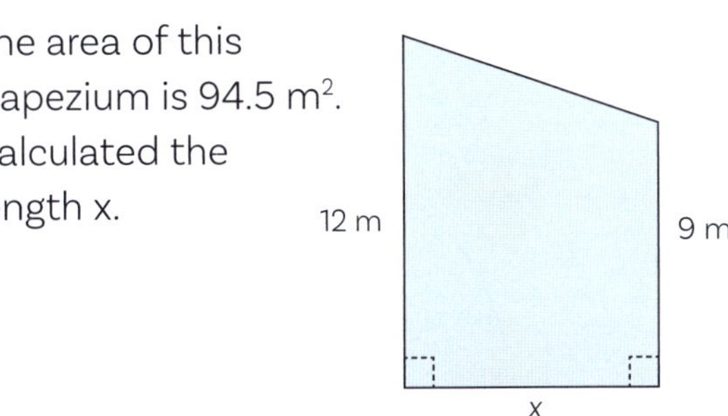

2 Circles

Area of a circle = πr^2

Example 1:

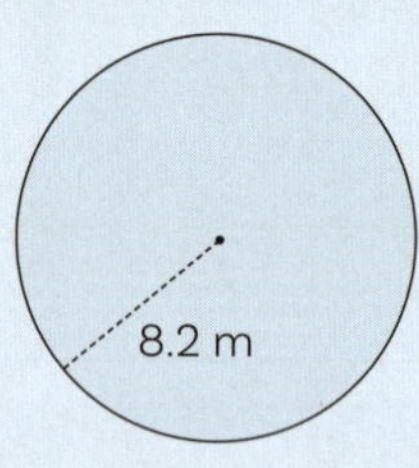

Area = π x 8.2^2
= 211.24 m^2

Example 2:

If you are given the diameter, you will need to **halve** it to find the radius.

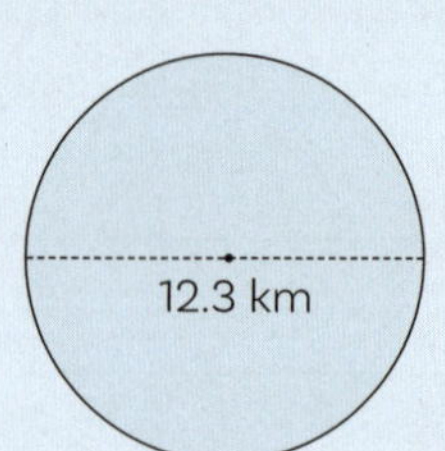

Area = π x 6.15^2
= 118.82 km^2

Calculate the shaded area in the diagrams. Round your answers to 4 sf.

1

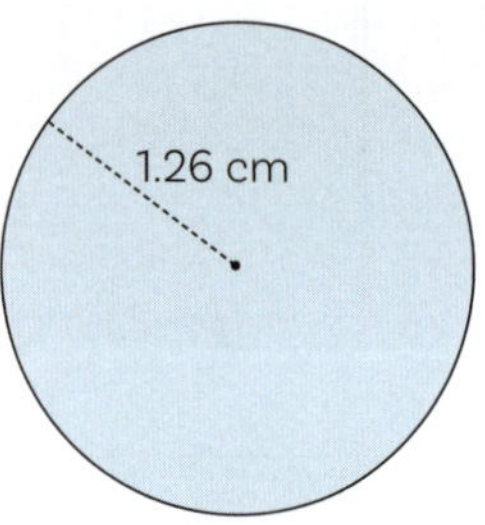

2

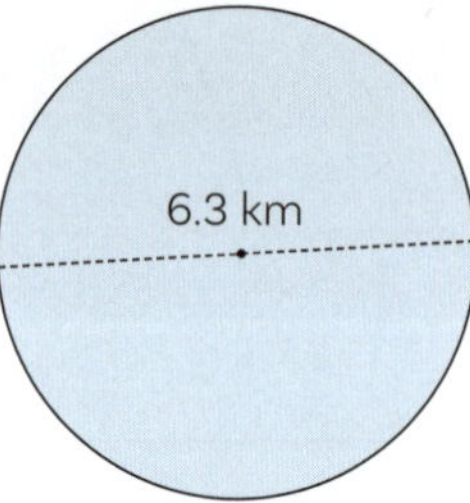

3

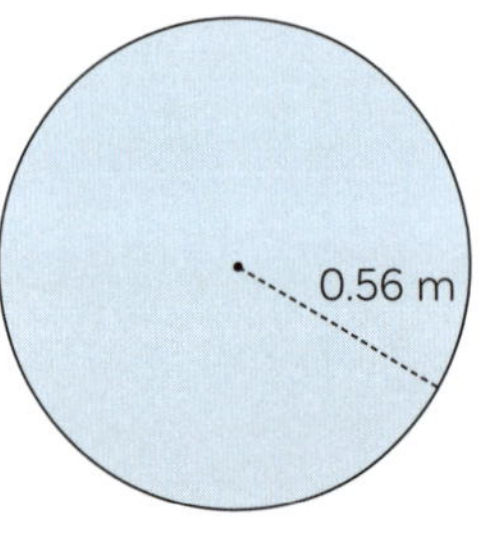

4

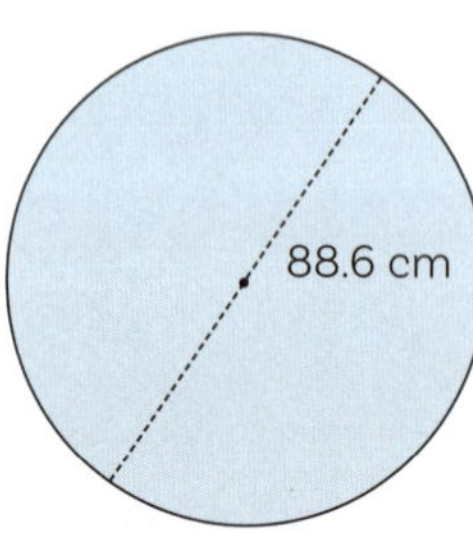

5

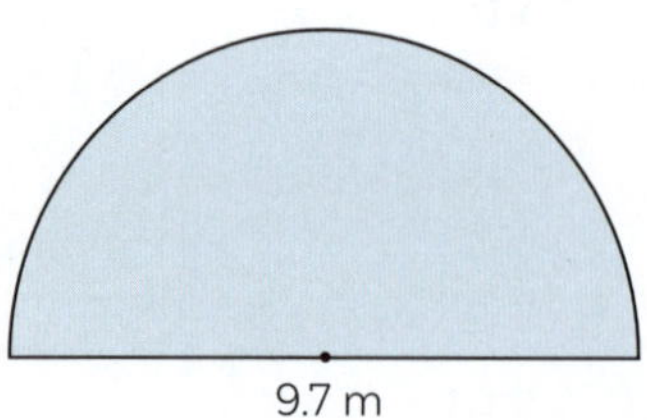

6

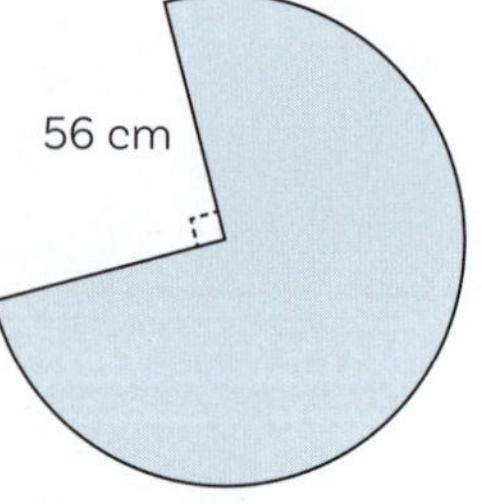

 ISBN: 9780170484084

7

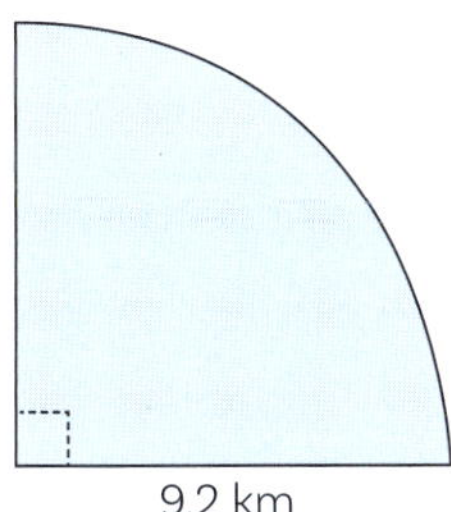

8 The area of this circle is 113.097 cm^2. Calculate its radius.

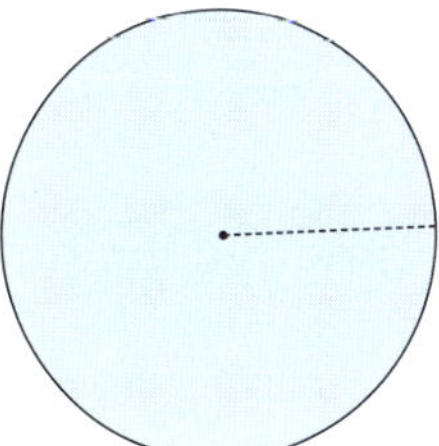

9 The area of this circle is 1134.11 m^2. Calculate its diameter.

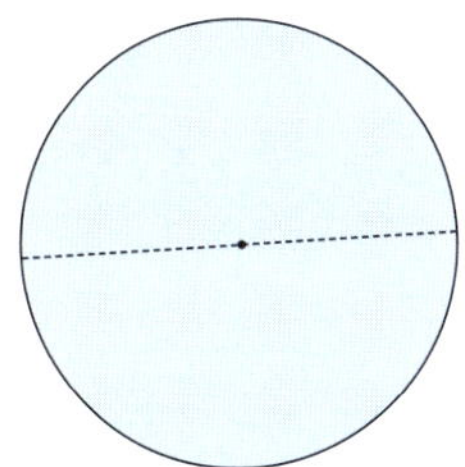

10

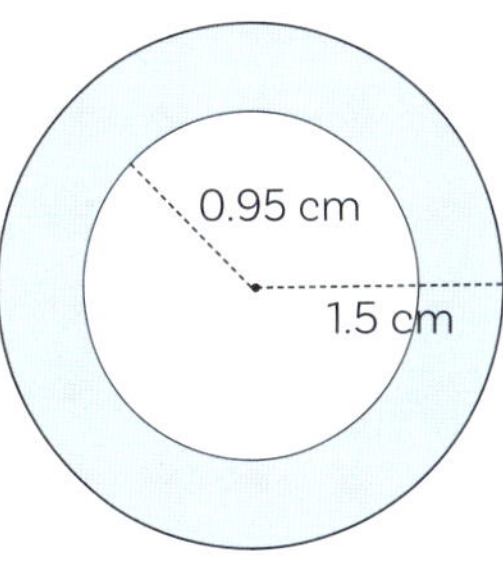

11 The area of this semicircle is 0.0831 m^2 Calculate its radius.

12

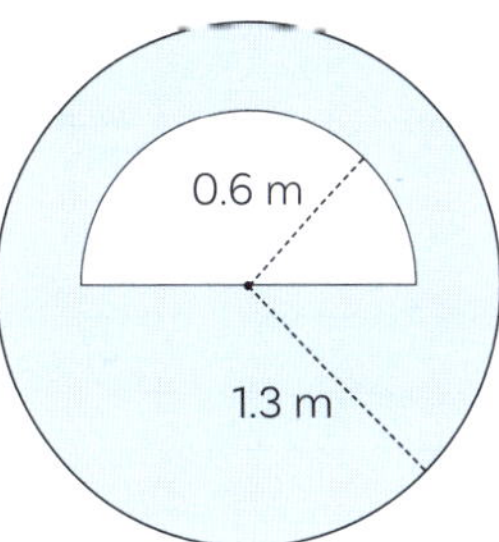

13 This shape is made from two semicircles with the same centre. Calculate the shaded area.

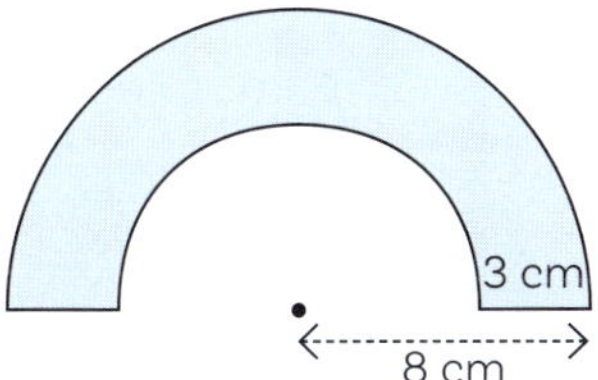

14 This figure is similar to that in **13**, but larger. Its area = 100.53 cm^2. Calculate x, the radius of the larger semicircle.

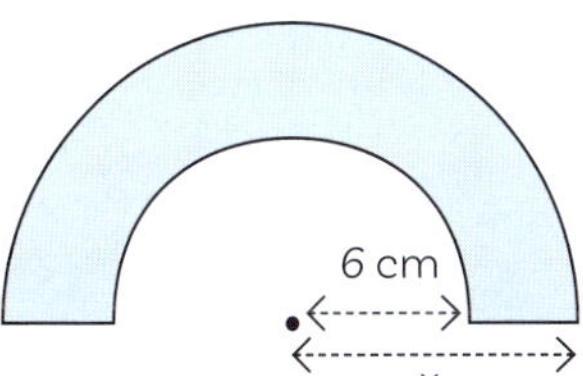

3 Compound shapes

- Compound shapes are shapes that are made up of other simple shapes.
- You need to find the areas of the separate shapes, and then add or subtract them.

Example: Sometimes you might be able to calculate the whole shape at once.

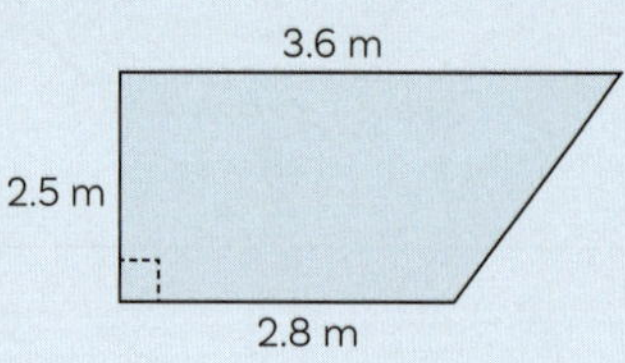

$$\text{Area} = \frac{a+b}{2} \times h$$
$$= \frac{3.6 + 2.8}{2} \times 2.5$$
$$= 8 \text{ m}^2$$

Or you can divide them into simpler shapes and then add the areas.

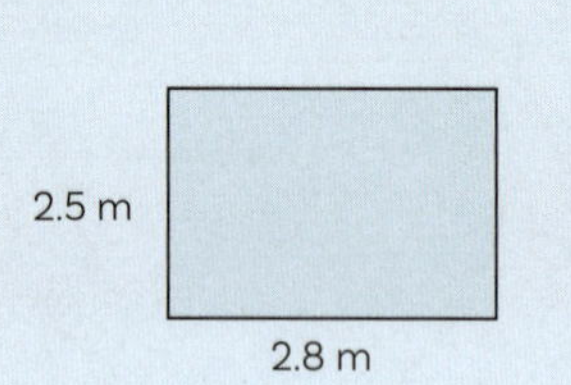

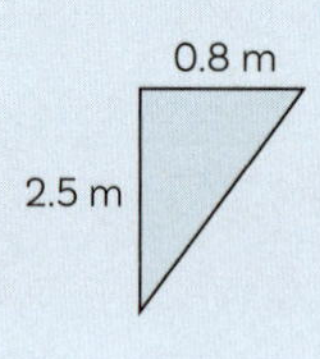

$$\text{Area} = b \times h + \frac{1}{2} \times h \times (a - b)$$
$$= 2.8 \times 2.5 + \frac{1}{2} \times 2.5 \times 0.8$$
$$= 7 + 1$$
$$= 8 \text{ m}^2$$

Calculate the areas of the shaded shapes. Round your answers to 4 sf.

1

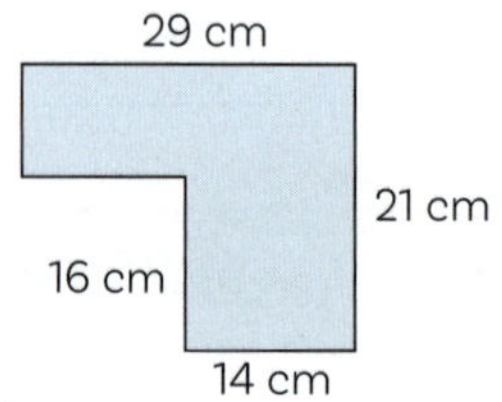

2

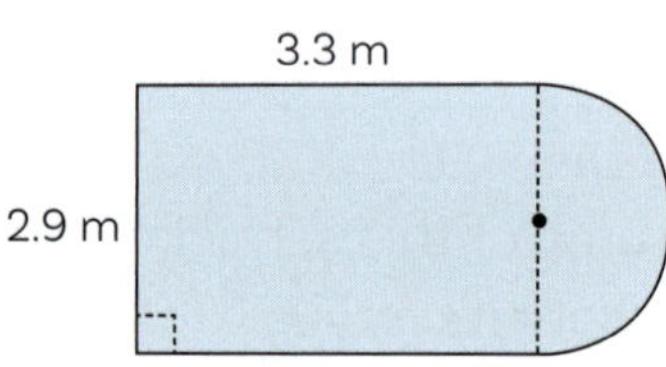

3

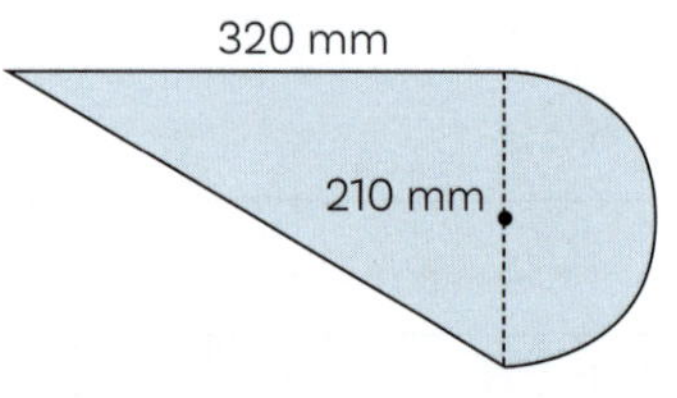

4

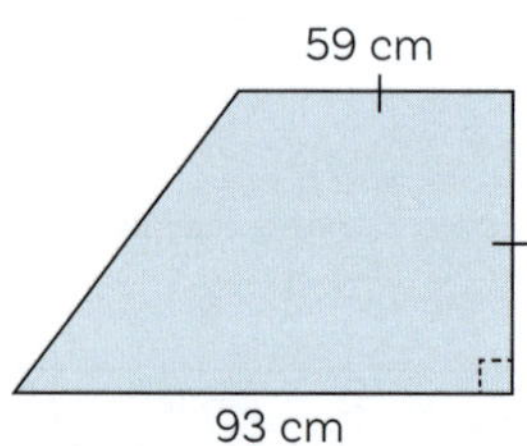

5

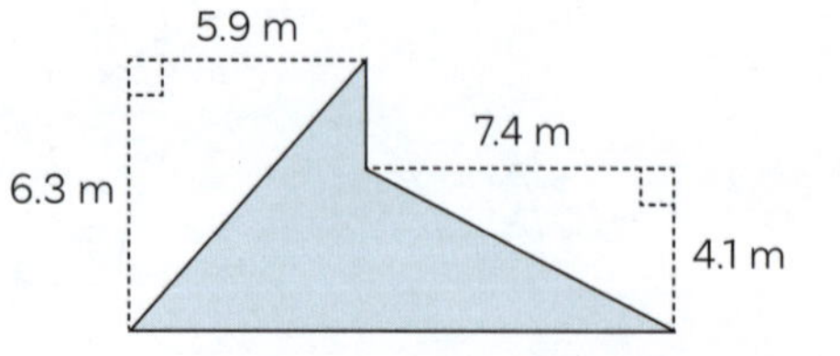

6

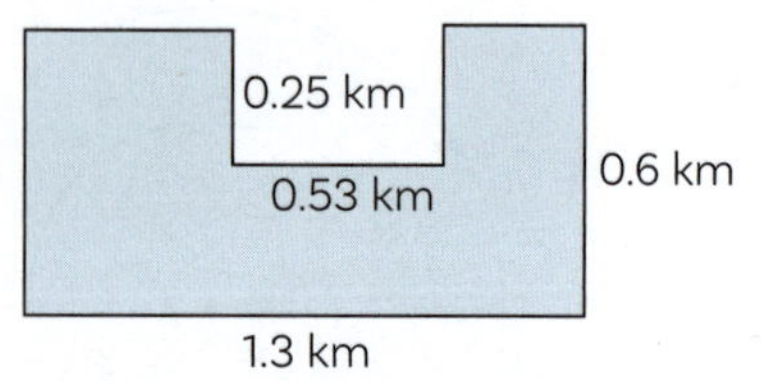

ISBN: 9780170484084

4 Shapes with holes

Calculate the area of the entire shape and subtract the area of the hole.

Example:

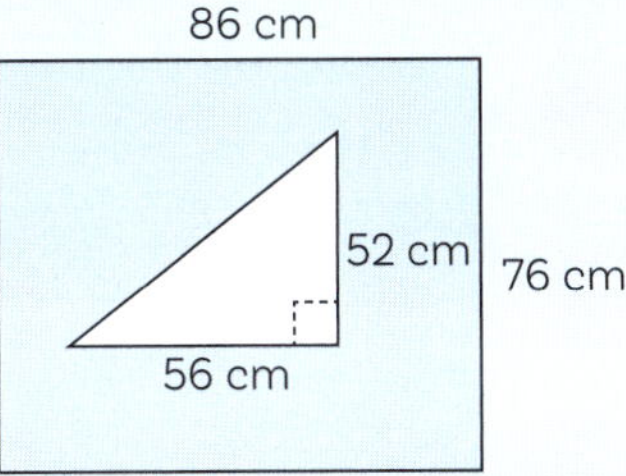

Shaded area = area rectangle - area triangle

$= (76 \times 86) - (0.5 \times 52 \times 56)$

$= 6536 - 1456$

$= 5080 \text{ cm}^2$

Calculate the area of the shaded shapes.

1 The larger shape is a rectangle.

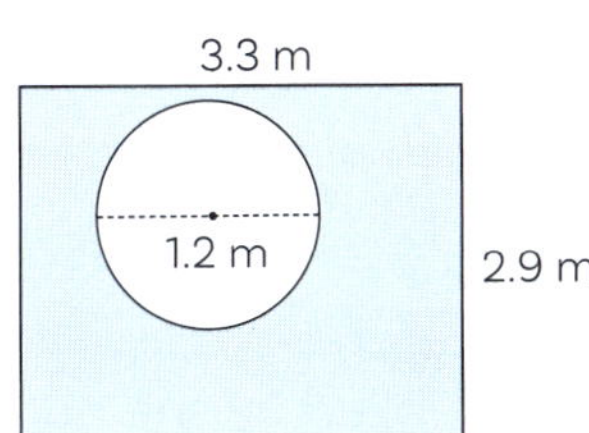

2

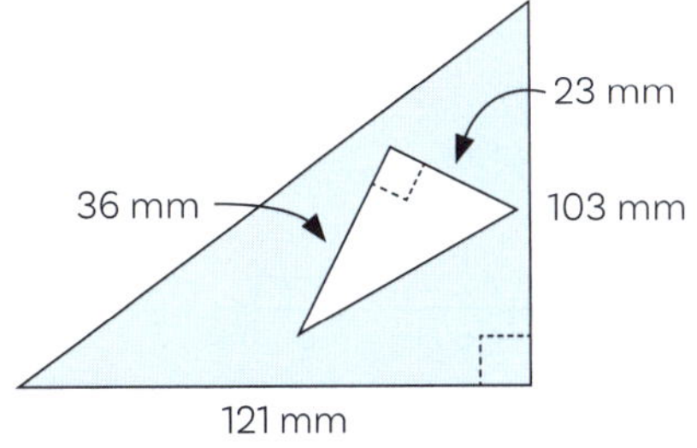

3

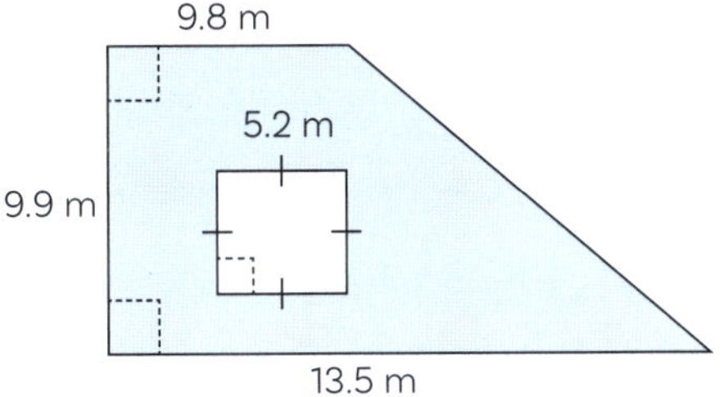

4

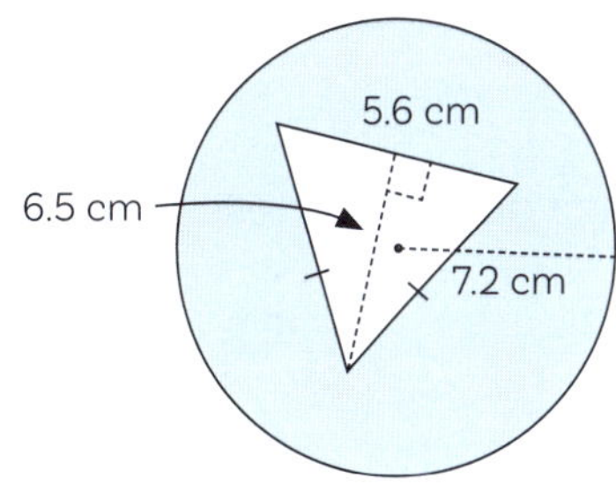

5

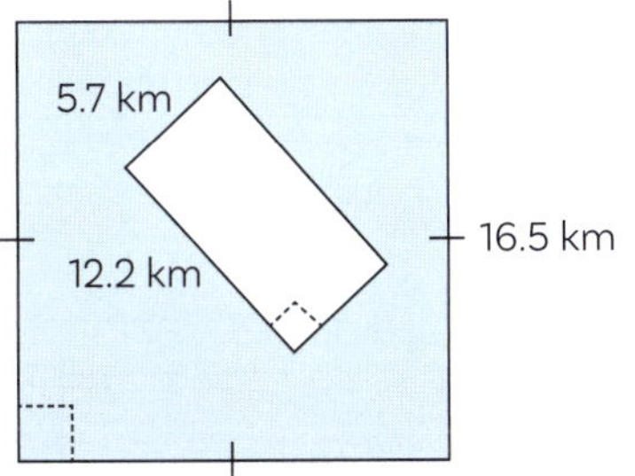

6

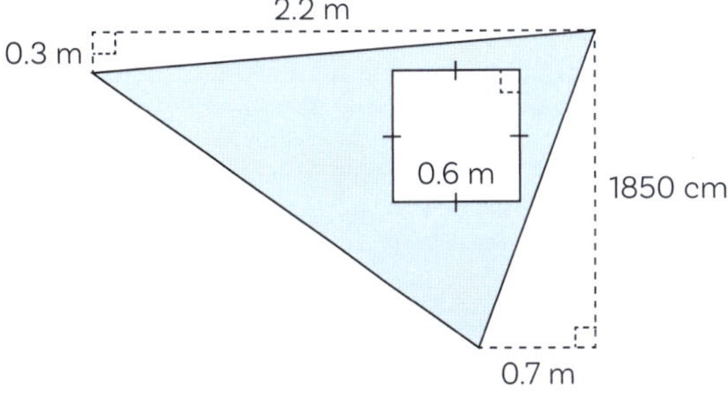

Volume

1 Cuboids

Volume = height x width x depth
= *h* x *w* x *d*

- The volume of a 3D shape is the size of the space inside the shape.
- A cuboid is a box shape.
- A cube is a box shape for which all the dimensions are equal.
- To find the volume of a cube or cuboid, multiply the **height** by the **width** by the **depth** (in any order).

Example 1:

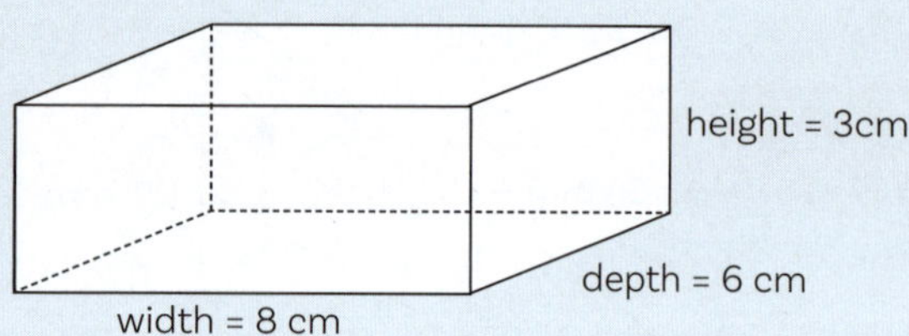

Volume = 3 cm x 8 cm x 6 cm
= 144 cm^3

When calculating volume, the units have ^{3} after them.

Example 2: Another way to think of it is the area of the 'face' multiplied by the depth.

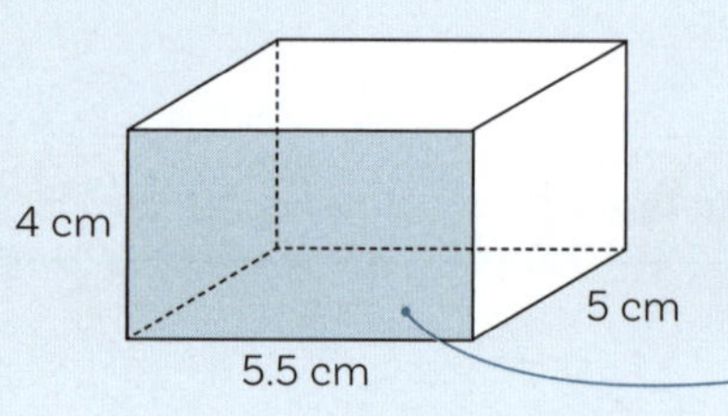

Volume = (4 x 5.5) x 5
= (22) x 5
= 110 cm^3

Area of the face is 22 cm^2, then multiply by the depth.

Calculate the volume of these shapes.

1

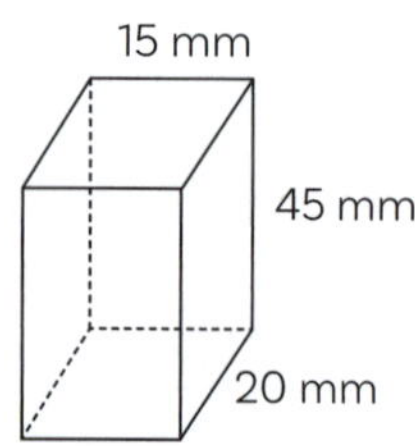

2

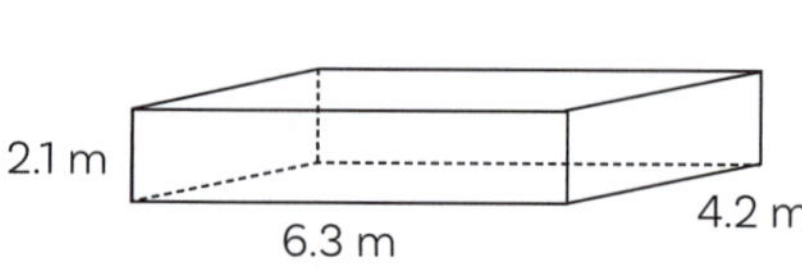

3

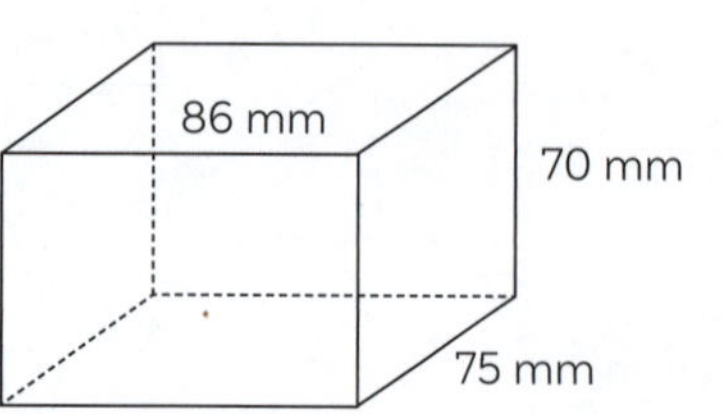

4

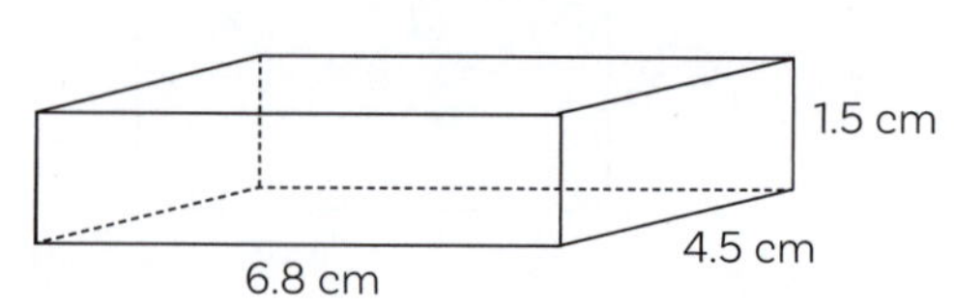

ISBN: 9780170484084

2 Cylinders

Volume = area of face x length

$= \pi r^2 h$

For the volume of a cylinder, find the area of the face and multiply it by the **height** (or **length**).

Example 1:

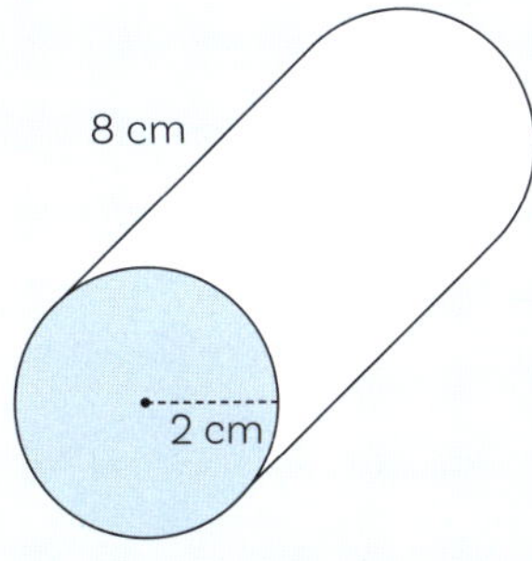

Volume = area of circle x length

$= \pi \times 2^2 \times 8$

$= 100.5\ cm^3$

Remember to round appropriately.

Example 2: Cylinders could be vertical. Again, find the area of the face and multiply by the height.

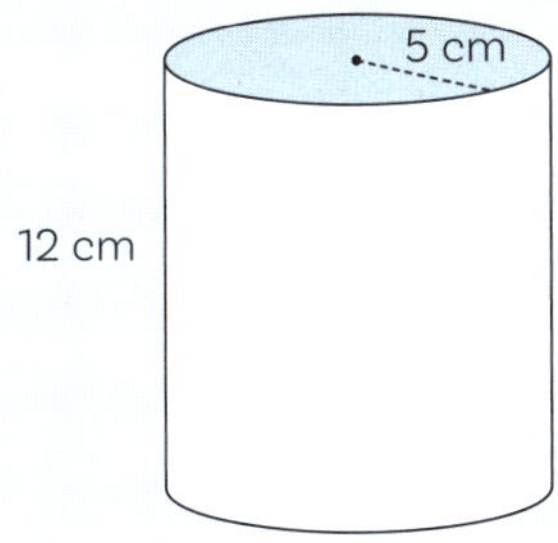

Volume = area of circle x height

$= \pi \times 5^2 \times 12$

$= 942.5\ cm^3$

Calculate the volume of these shapes. Round your answers to 4 sf.

1

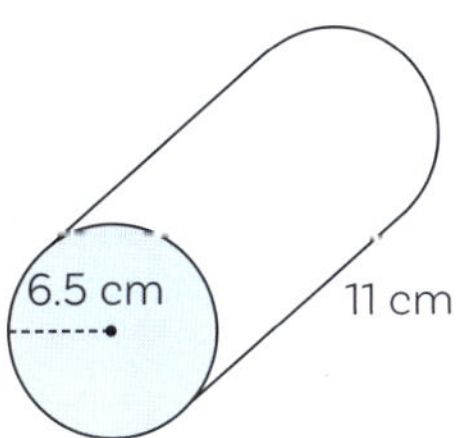

2

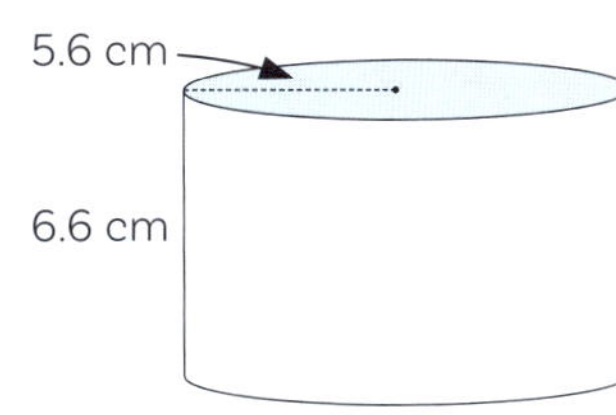

3

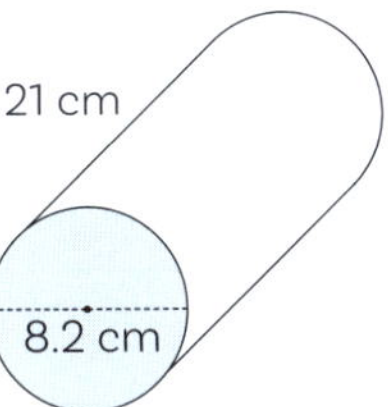

4

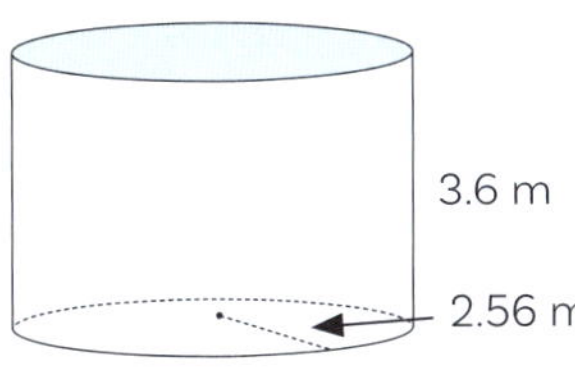

5

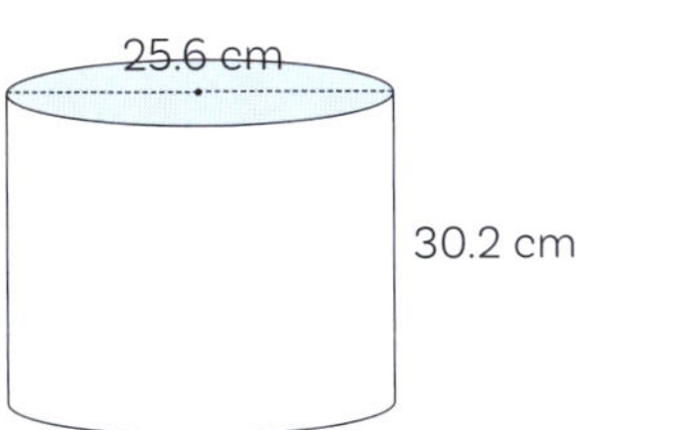

6

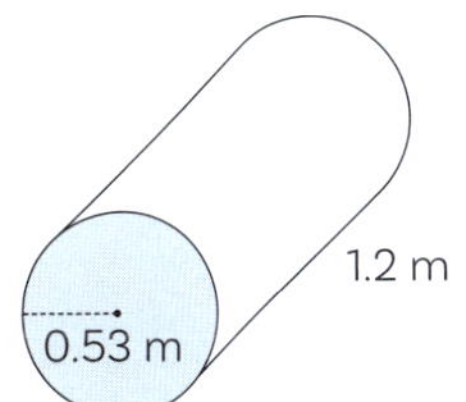

ISBN: 9780170484084

3 Triangular prisms

$$\textbf{Volume} = \textbf{area of face} \times \textbf{length}$$
$$= \frac{1}{2}bh \times l$$

For the volume of a triangular prism, find the area of the face and multiply it by the **length** (or **height**).

Example 1:

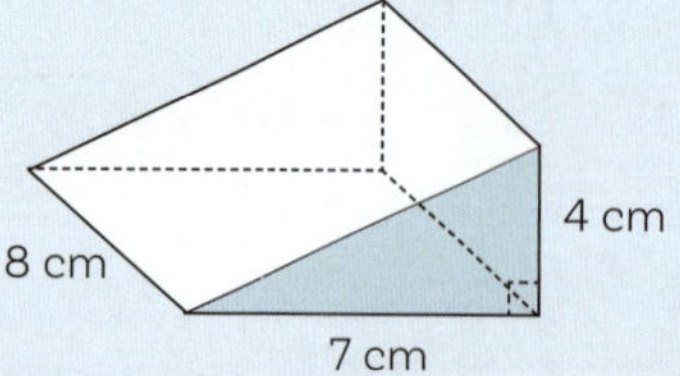

Volume = $(\frac{1}{2} \times 7 \times 4) \times 8$

= 112 cm^3

Remember, the area of a triangle = $\frac{1}{2}$ x base x height.

Example 2:

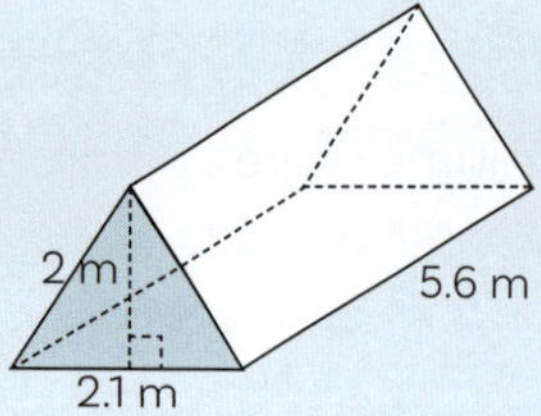

Volume = $(\frac{1}{2} \times 2.1 \times 2) \times 5.6$

= 11.76 m^3

Calculate the volume of these shapes. Round your answers to 4 sf.

1

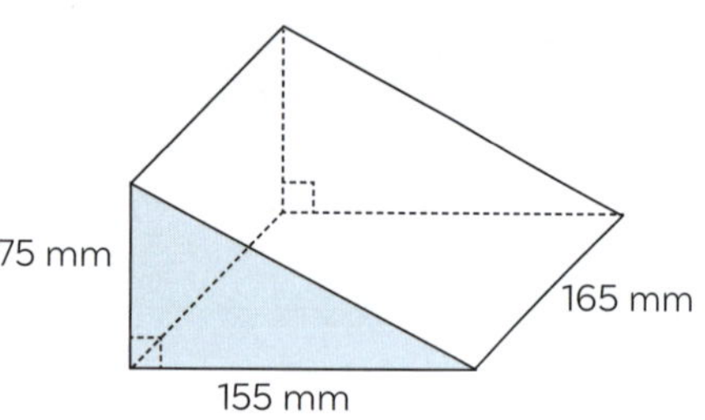

2

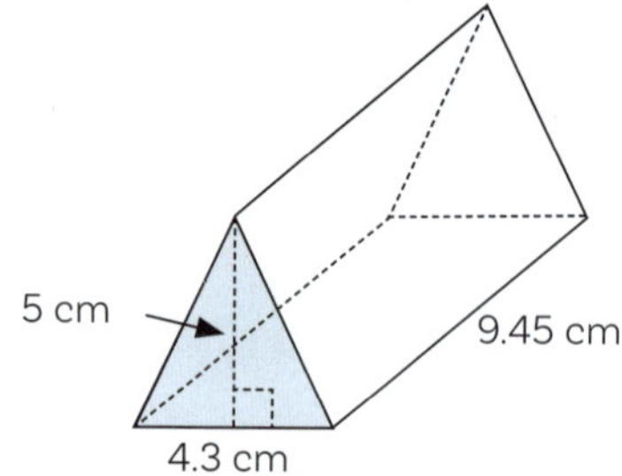

3

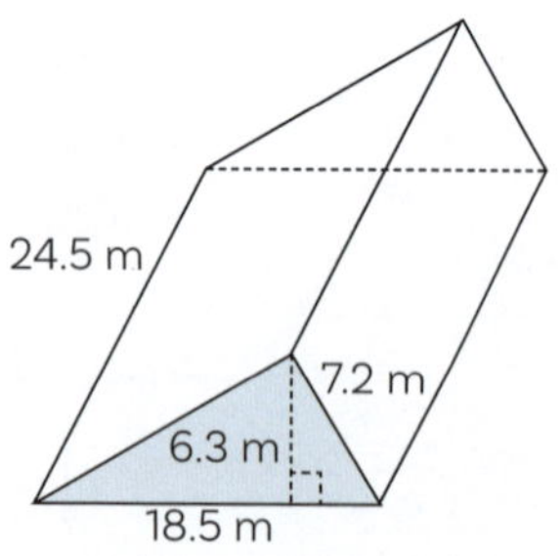

4

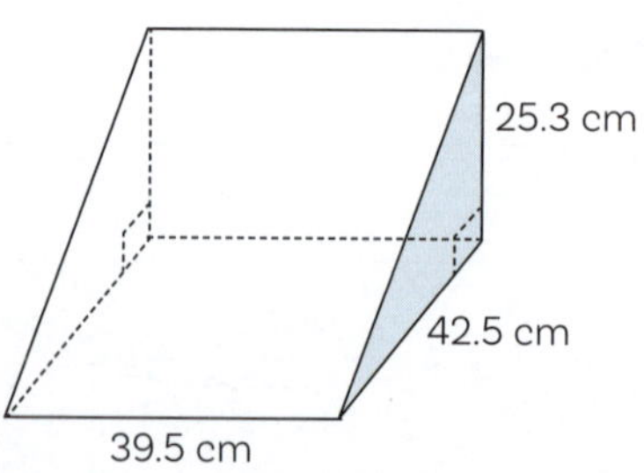

 ISBN: 9780170484084

Surface area

1 Cuboids

- The surface area of a 3D shape is the total area of the outside of the shape.
- To find the surface area of any shape, find the areas of all the sides and add them together.
- Drawing the net of the shape is very helpful.

Example 1:

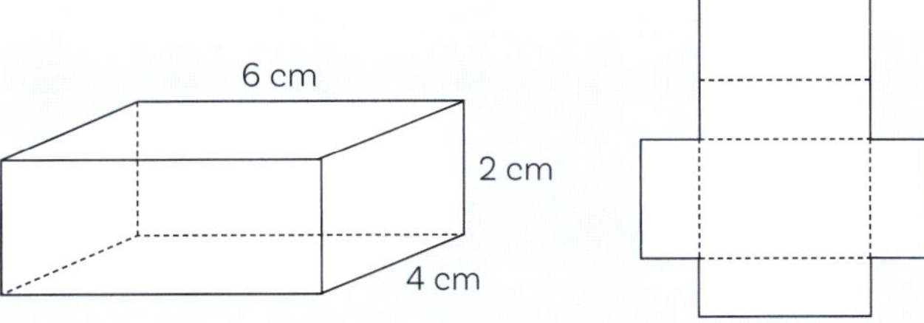

Surface area = (6 x 4 + 4 x 2 + 6 x 2) x 2
= 88 cm^2

There are two of each of these sides.

Example 2:

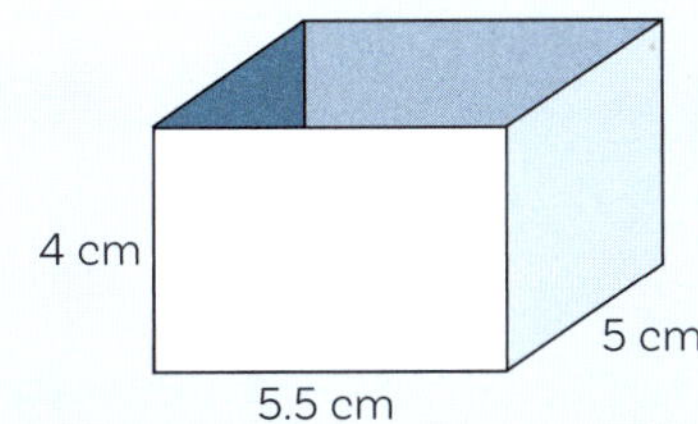

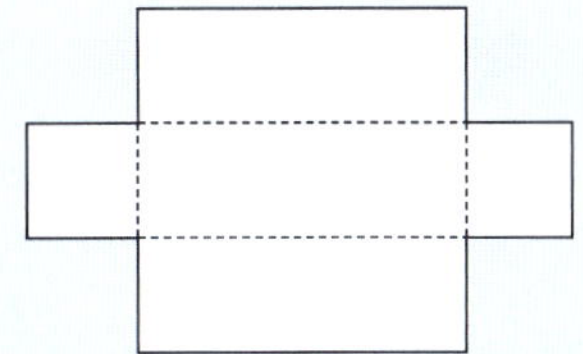

Surface area = (4 x 5.5 + 4 x 5) x 2 + (5.5 x 5)
= 111.5 cm^2

No lid, so only one 5.5 x 5 side.

Calculate the surface area of these shapes.

1

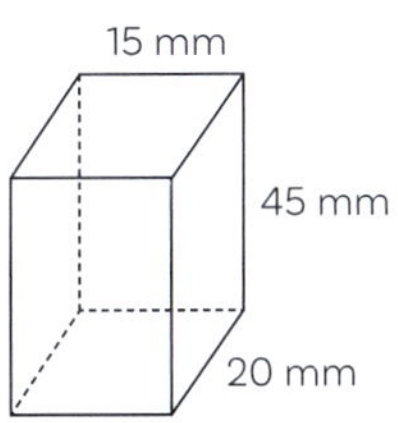

2

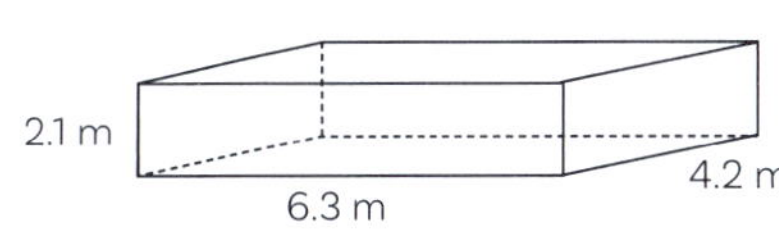

3

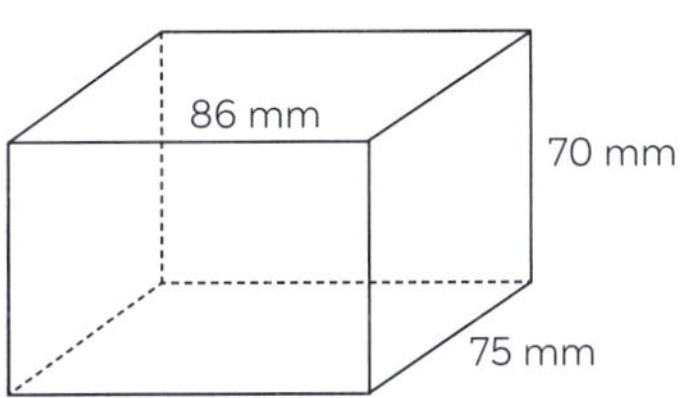

4

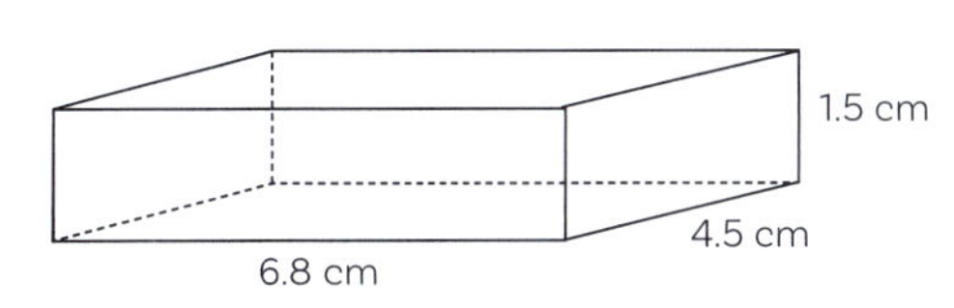

ISBN: 9780170484084

2 Cylinders, prisms and pyramids

Again, it is useful to draw the net.

Cylinder

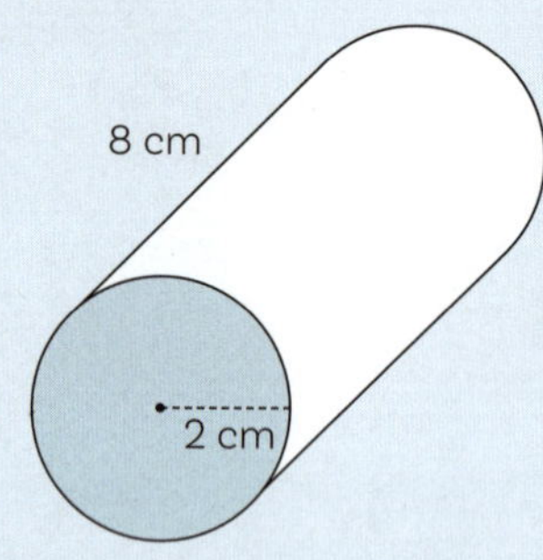

Net

$$\text{Surface area} = 2(\pi \times 2^2) + (8 \times \pi \times 4)$$
$$= (25.1327) + (100.5310)$$
$$= 125.66 \text{ cm}^2$$

Rectangle (the circumference is the base).

Prism

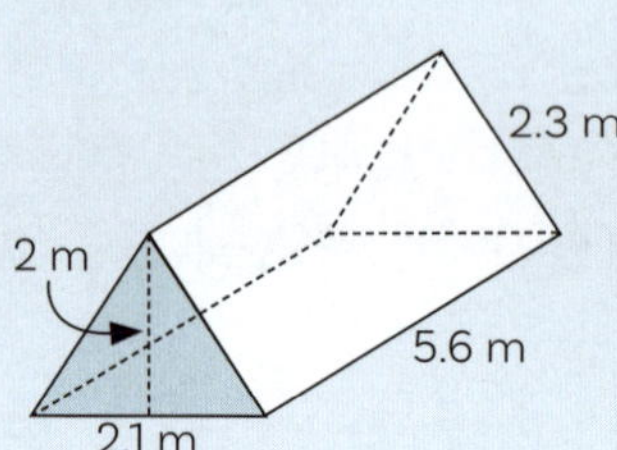

Net

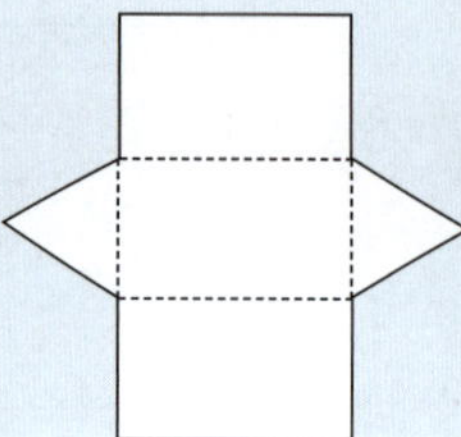

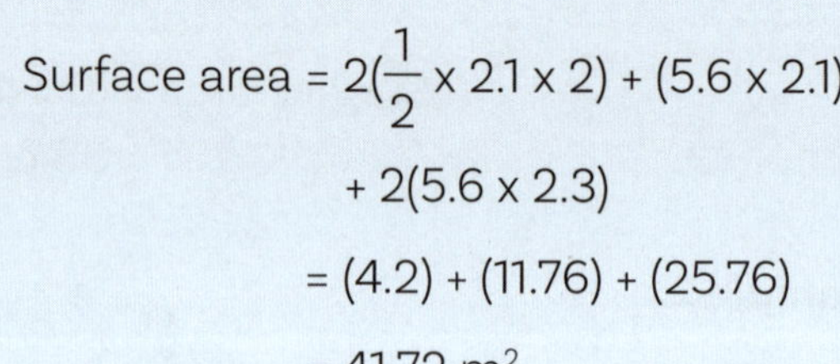

$$\text{Surface area} = 2(\frac{1}{2} \times 2.1 \times 2) + (5.6 \times 2.1)$$
$$+ 2(5.6 \times 2.3)$$
$$= (4.2) + (11.76) + (25.76)$$
$$= 41.72 \text{ m}^2$$

Pyramid

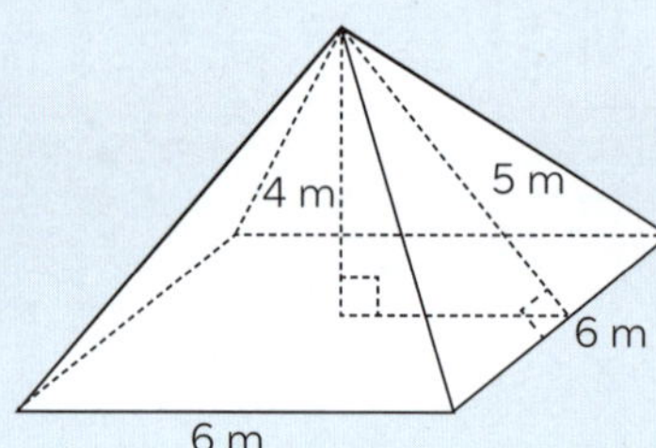

Net

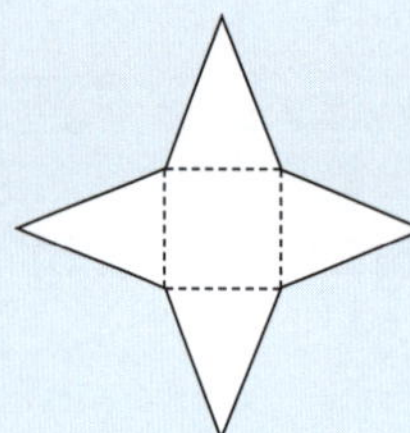

$$\text{Surface area} = (6 \times 6) + 4(0.5 \times 6 \times 5)$$
$$= (36) + (60)$$
$$= 96 \text{ m}^2$$

Calculate the surface areas of the following shapes.

1

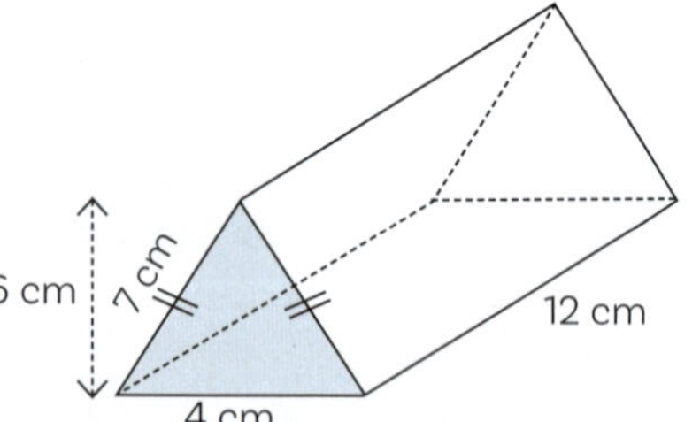

Net

____

____

____

____

____

2

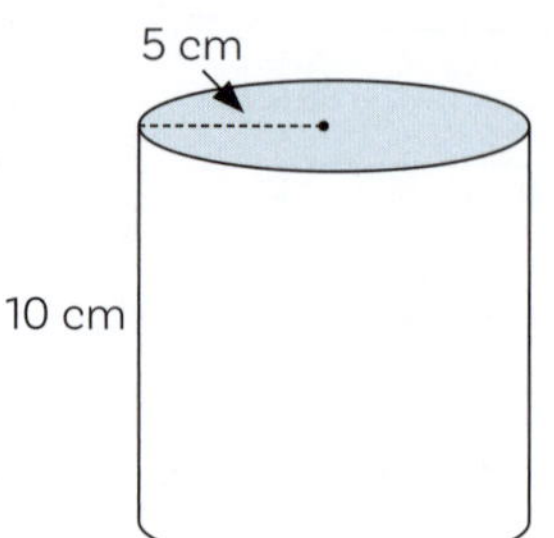

Net

____

____

____

____

____

 ISBN: 9780170484084

3 This pyramid is on a square base.

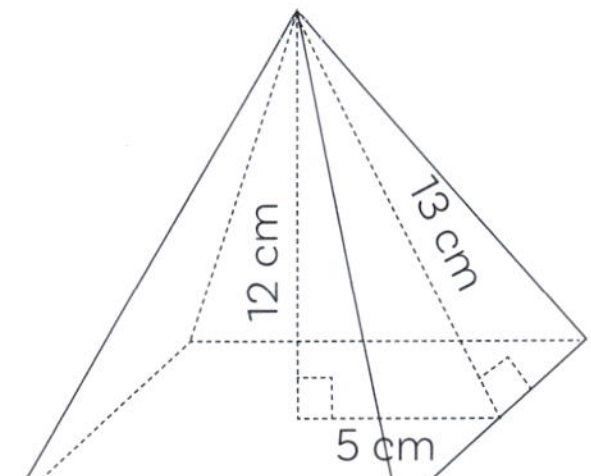

Net

4

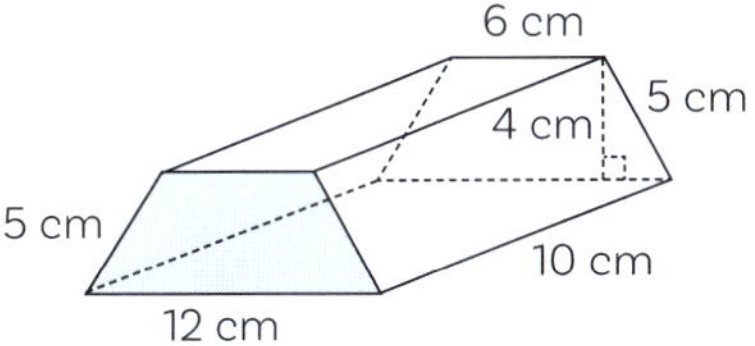

Net

5

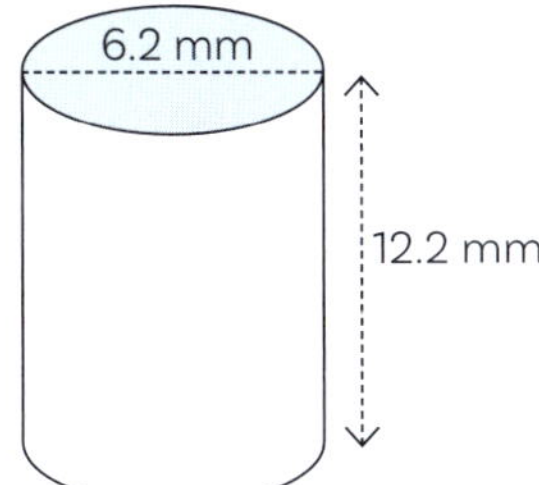

Net

6

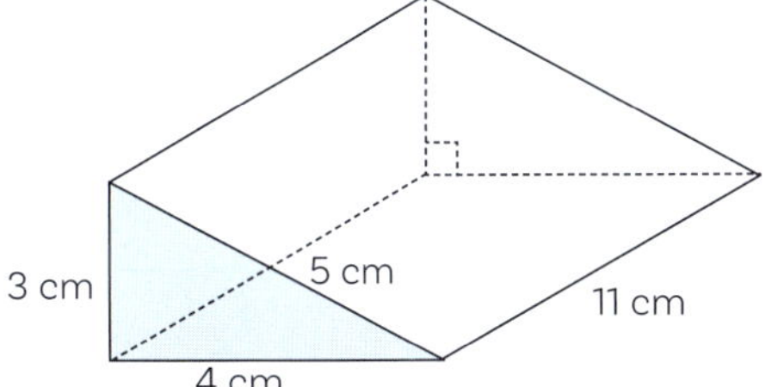

Net

7 An unsharpened 20 cm long pencil with the dimensions shown.

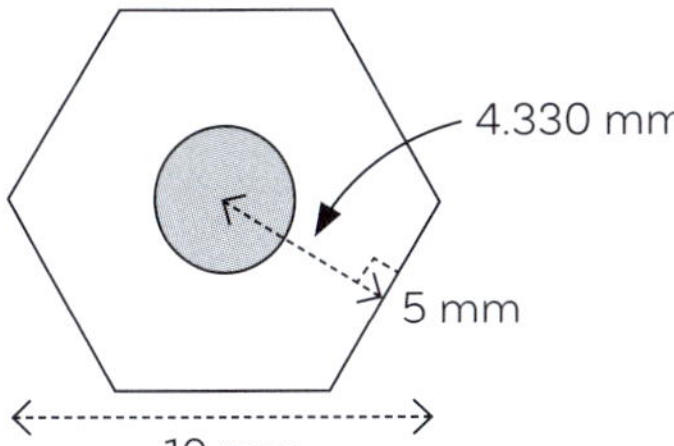

Net

ISBN: 9780170484084

STATISTICS AND PROBABILITY

Probability

The range of values for probabilities

0	0.5	1
Impossible	50:50 chance	Certain

Discuss the meanings of the following words and phrases with others in your class, and match each to its appropriate probability value(s). You may have several words beside the same probability value, and some words can cover several probability values.

very unlikely, improbable, slight chance, a sure thing, extremely likely, maybe, likely, impossible, possible, probable, unlikely, no chance, definite, no way

0	
0.1	
0.2	
0.3	
0.4	
0.5	
0.6	
0.7	
0.8	
0.9	
1.0	

ISBN: 9780170484084

Using numbers for writing probability

- Probabilities can be written as **fractions, decimals or percentages**.
- You can convert between these with your calculator.
- When you want to compare probabilities, you should always use **decimals**.

Example: Write the probability $\frac{7}{16}$ as a decimal. 7 [÷] 16 [=] 0.4375

Convert the following probabilities to decimals, and state which of each pair is more likely.

1 $\frac{7}{16}$ = ____________ $\frac{1}{3}$ = ____________

More likely: ____________________

2 $\frac{3}{4}$ = ____________ $\frac{8}{11}$ = ____________

More likely: ____________________

3 $\frac{17}{18}$ = ____________ $\frac{19}{20}$ = ____________

More likely: ____________________

4 $\frac{13}{24}$ = ____________ $\frac{27}{50}$ = ____________

More likely: ____________________

5 $\frac{11}{20}$ = ____________ $\frac{7}{13}$ = ____________

More likely: ____________________

6 $\frac{6}{17}$ = ____________ $\frac{3}{8}$ = ____________

More likely: ____________________

Test yourself

For the following situations, indicate whether each is right or wrong, and justify your decision.

✔ 

7 Nikki said that the probability is 0.009 that tomorrow will be a sunny day.

____

8 Adam calculated that the probability was -0.35 that he would win a game.

____

9 Tahu calculated that the probability that he would pass Level 1 was 105%.

____

10 Sora said that the probability was $\frac{1}{200}$ that she would be late for school.

____

ISBN: 9780170484084

Ways of calculating probabilities

$$\text{Probability} = \frac{\text{number of favourable outcomes}}{\text{total possible outcomes}}$$

Example 1: For a fair die, P(1 or a 2) = $\frac{2}{6}$ or $\frac{1}{3}$ or $0.\dot{3}$

Example 2: Using a pack of cards with no jokers (52 cards):

P(spade) = $\frac{13}{52}$ or $\frac{1}{4}$ or 0.25

P(ace of spades) = $\frac{1}{52}$ or 0.01923

P(ace) = $\frac{4}{52}$ or $\frac{1}{13}$ or 0.0769

P(spade or ace) = $\frac{16}{52}$ or 0.3077

Calculate the following probabilities.

When tossing a fair die:

1 P(6) = ____________________

2 P(4, 5 or 6) = ____________________

3 P(odd number) = ____________________

4 P(not a 6) = ____________________

5 P(7) = ____________________

6 P(1, 2, 3, 4, 5 or 6) = ____________________

Using a pack of cards with no jokers (52 cards), and selecting one card at random:

7 P(club) = ____________________

8 P(black card) = ____________________

9 P(10) = ____________________

10 P(black 10) = ____________________

11 P(5 or 6) = ____________________

12 P(red 5, red 6 or red 7) = ____________________

13 P(not an ace) = ____________________

14 P(not a king or queen) = ____________________

Mike has a bag of lollies. It contains 7 red lollies, 3 green ones, 5 yellow ones and 9 white ones. He puts his hand in the bag and selects a lolly at random:

15 P(red) = ____________________

16 P(not getting a red) = ____________________

17 P(red or green) = ____________________

18 P(not getting a red or green) = ____________

19 P(white, red or green) = ____________________

20 P(not yellow) = ____________________

21 P(black) = ____________________

22 P(red, green, yellow or white) = ____________

ISBN: 9780170484084

Expected number of outcomes

Expected number of outcomes = P(event) x number of trials

Example 1: Amy tosses a die 120 times. How many times can she expect a 6?

Expected number of 6s = P(6) x 120

$= 0.1\dot{6} \times 120$

= 20

Example 2: The probability of a person being struck by lightning in the USA is $\frac{1}{700\,000}$ in one year. If in 2024 the population of the USA is about 341 814 000, calculate the expected number of people who will be struck by lightning during 2024.

Expected number of people struck by lightning = $\frac{1}{700\,000} \times 341\,814\,000$

= 488 or 489

The exact answer is 488.3...., but this is people, so we must round down or up to the nearest whole person.

Answer the following questions.

1 Archie flips a coin 80 times and records the results. How many 'heads' can he expect?

____

2 Nick tossed a die 90 times and recorded the results. How many times can he expect to get a 5 or a 6?

____

3 The probability is 0.1 that the person will be left-handed. If your school roll is 793, how many people would you expect to be left-handed at your school?

____

4 The probability is 0.08 that the boy will be colour blind. If there are 384 boys at your school, how many would you expect to be colour blind?

____

5 The probability is 0.001 that an egg will have two yolks. If a farm produces 150 dozen eggs each week, how many would you expect to have two yolks?

____

6 In the game of poker, each player is dealt a hand of five cards. A full house occurs when three of the cards have the same number, and the other two also have the same number, for example 6, 6, 6, 2, 2. The probability of a full house is 0.001441. If 720 hands are dealt during the course of a series of games, how many of these would be expected to contain a full house?

____

Combining probabilities — probability trees

Probability trees are very useful for calculating probabilities where several events occur.

Example: Angus had a bag of lollies. It contained 6 red ones and 4 green ones. Without looking, he selected a lolly and ate it. Then he selected a second lolly and ate it.

a What is the probability that he ate two red ones?

b What is the probability that he ate one of each colour?

Steps:

1 Decide what the **events** are, and their order:
Event 1 will be the colour of the first lolly.
Event 2 will be the colour of the second lolly.
Write the **events** at the **ends** of the branches:

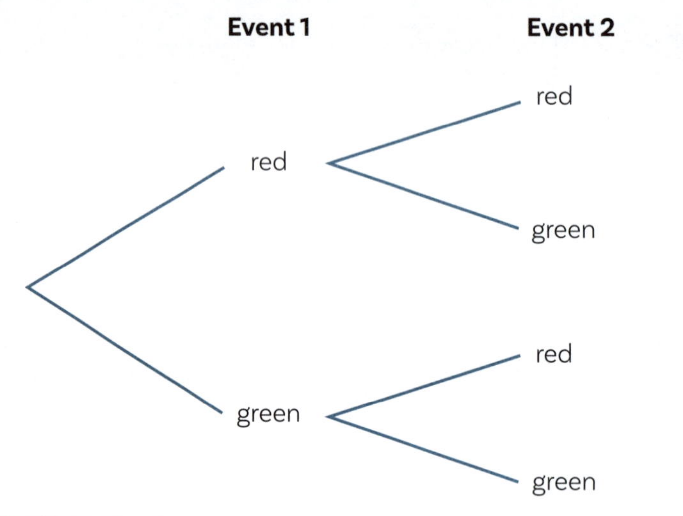

2 Add the **probabilities** of each event to the **middle** of each branch:

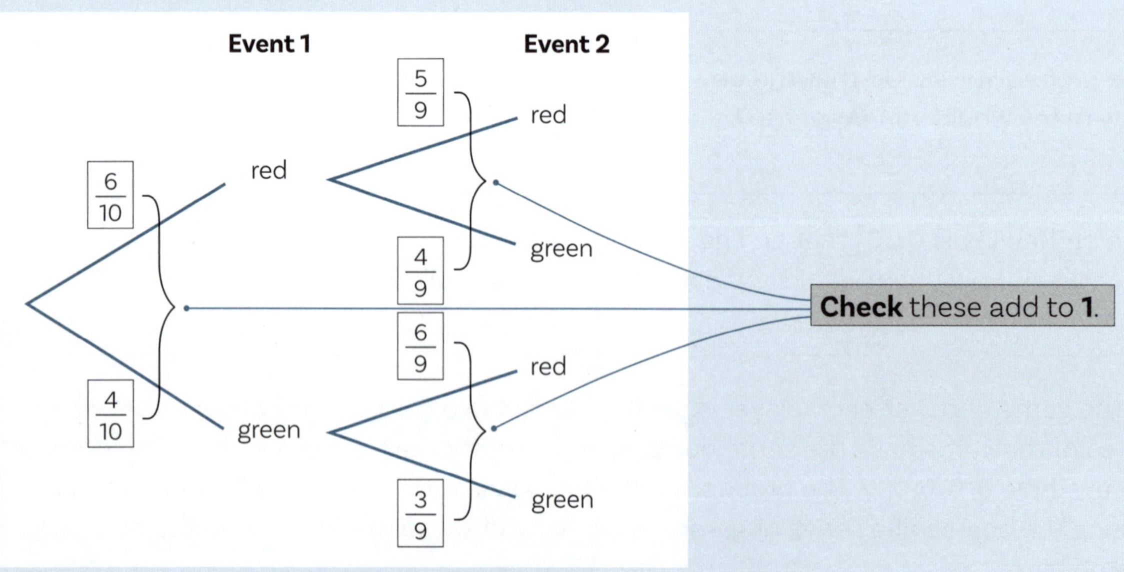

3 **Check** that the probabilities for every branch add to **1**.

ISBN: 9780170484084

4 **List** the outcomes at the ends of each branch, and calculate the probabilities at each end. You **multiply** the probabilities along each branch because Event 1 **and** Event 2 must occur.

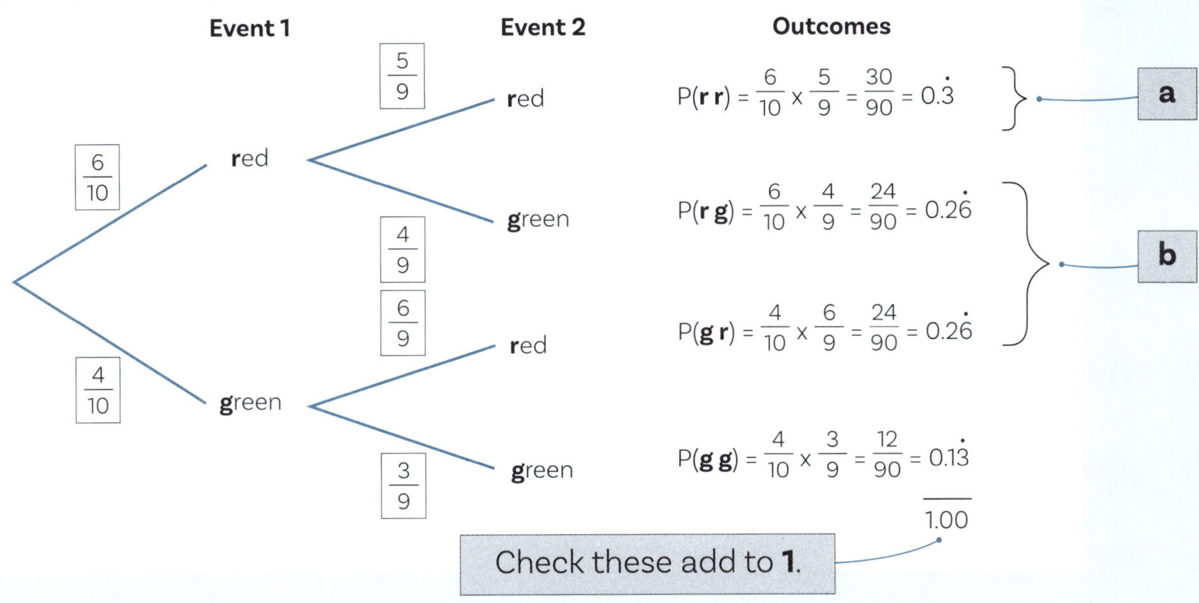

5 **Check** that your probabilities in the right-hand column **add** to **1**. This is because **one** of these options **must** occur.

6 In the Outcomes column, highlight the event(s) required, along with their probabilities. **Add** these to find the overall probability required.

a $P(rr) = 0.\dot{3}$

b $P(rg \text{ or } gr) = 0.2\dot{6} + 0.2\dot{6} = 0.5\dot{3}$

Useful tips:

1 **Addition Rule:** If one event **OR** another happens ⇒ **ADD** the probabilities.

2 **Multiplication Rule:** If one event **AND** another happens ⇒ **MULTIPLY** the probabilities.

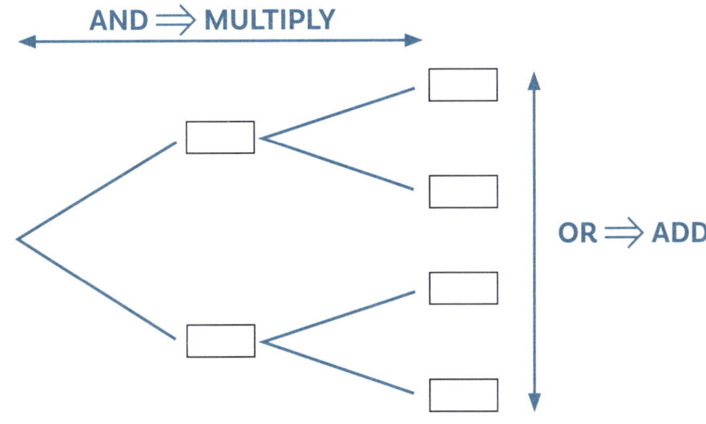

ISBN: 9780170484084

Use probability trees to answer the following questions.

1 Chris had a bag of lollies. He had 5 yellow ones and 3 blue ones. Without looking, he selected a lolly and then put it back in the bag. He then selected a second lolly.

a Complete the probability tree.

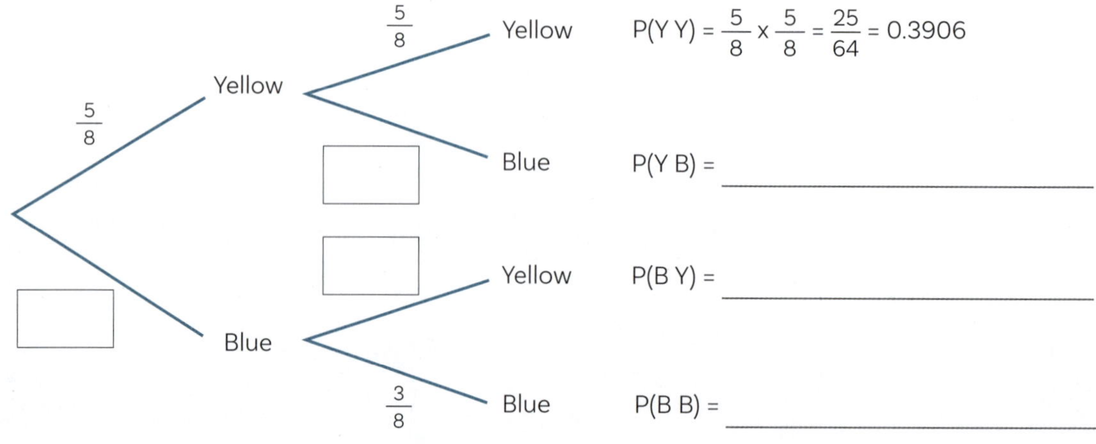

b What is the probability that he selected two yellow ones?

____

c What is the probability that he selected one of each colour?

____

d What is the probability that he selected two of the same colour?

____

e What is the probability that he doesn't select a yellow lolly?

____

f What is the probability that he selects at least one blue lolly?

____

g If he repeated this experiment 200 times (without eating any), how many times could he expect to get two yellow lollies?

____

h If he repeated this experiment 200 times (without eating any), how many times could he expect to get at least one blue lolly?

____

ISBN: 9780170484084

2 Charlie had a bag of lollies. He had 5 yellow ones and 3 blue ones. Without looking, he selected a lolly and ate it. He selected and ate a second lolly as well.

a Complete the probability tree.

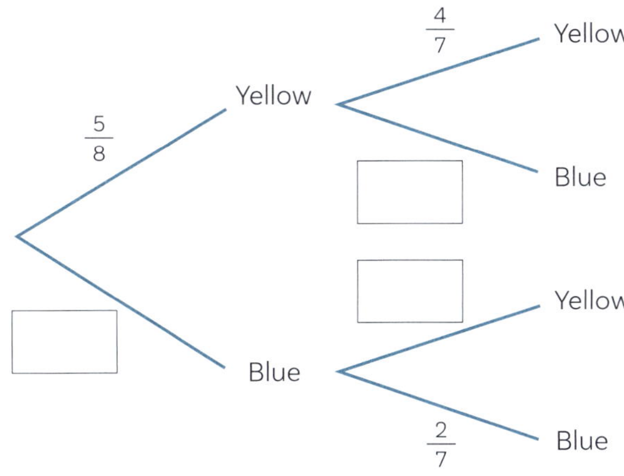

$P(Y\ Y) = \frac{5}{8} \times \frac{4}{7} = \frac{20}{56} = 0.3571$

P(Y B) = ____________________

P(B Y) = ____________________

P(B B) = ____________________

b What is the probability that he ate two yellow ones?

____

c What is the probability that he ate one of each colour?

____

d What is the probability that he ate two of the same colour?

____

e What is the probability that he doesn't eat a yellow lolly?

____

f What is the probability that he ate at least one blue lolly?

____

g What is the probability that after eating his two lollies, there are exactly two blue lollies left in the bag?

____

h What is the probability that after eating his two lollies, there are at least three yellow lollies left in the bag?

____

ISBN: 9780170484084

3 Sarah and three friends are planning their Saturday evening. Two of them want to have pizza and two want hamburgers. Three of them want to go to a movie, and one wants to watch videos. If they all have an equal say and all do the same things, answer the following.

a Complete the probability tree.

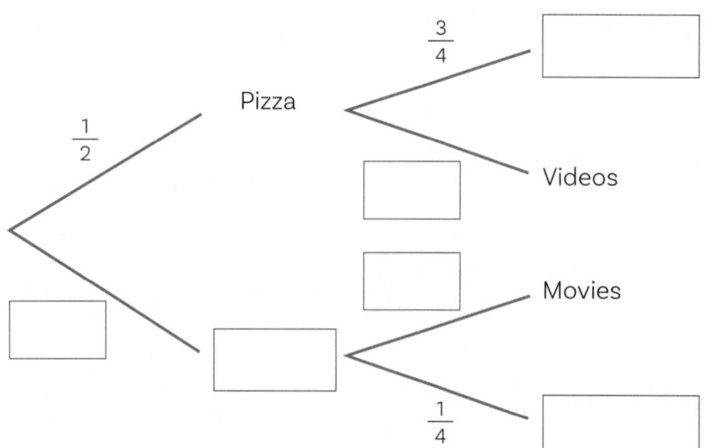

$P(P\ M) = \frac{1}{2} \times \frac{3}{4} = \frac{3}{8} = 0.375$

P(P V) = ______________________

P(H M) = ______________________

P(H V) = ______________________

b What is the probability that they go to a movie?

____

c What is the probability that they eat pizza and watch videos?

____

d What is the probability that they eat hamburgers and go to a movie?

____

e What is the probability that they either have pizza or go to a movie?

____

f What is the probability that they either have hamburgers or watch a video?

____

g What is the probability that they neither eat hamburgers nor watch videos?

____

h What is the probability that they neither eat hamburgers nor go to a movie?

____

ISBN: 9780170484084

Two-way tables

- When there is data on two different characteristics for each member of a population, you can display these in a two-way table.

Example: At a school of 434 students, each is asked about whether they have ever broken a bone and if they own a scooter. The results are shown in the table below.

	Broken a bone	Haven't broken a bone
Own a scooter	153	106
Don't own a scooter	78	97

It is a good idea to add totals to your table.

	Broken a bone	Haven't broken a bone	Totals
Own a scooter	153	106	**259**
Don't own a scooter	78	97	**175**
Totals	**231**	**203**	**434**

Check that both add to 434.

a What is the probability that a student has broken a bone?

$$P(\text{broken bone}) = \frac{231}{434} = 0.5323$$

b What is the probability that a student neither owns a scooter nor has broken a bone?

$$P(\text{neither}) = \frac{97}{434} = 0.2235$$

c If two students from this school are selected at random, what is the probability that both students have broken a bone?

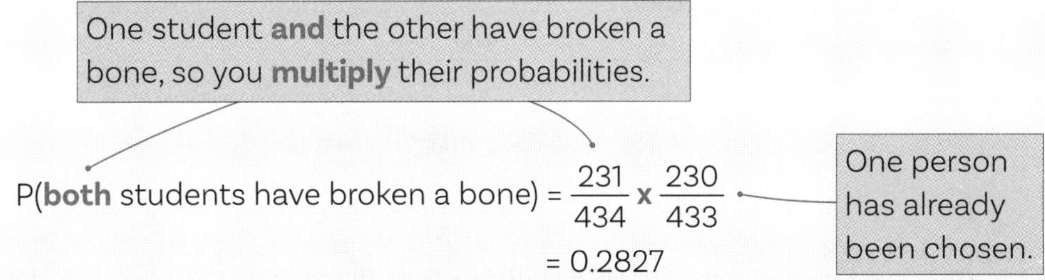

One student **and** the other have broken a bone, so you **multiply** their probabilities.

$$P(\textbf{both}\text{ students have broken a bone}) = \frac{231}{434} \times \frac{230}{433}$$

$$= 0.2827$$

One person has already been chosen.

d If a scooter owner is selected at random, what is the probability that they also have broken a bone?

$$P(\text{scooter owner has broken a bone}) = \frac{153}{259} = 0.5907$$

We are just concerned with the 259 scooter owners.

e Compare the answers to **a** and **d**. What do they tell you about scooter owners and broken bones at this school?

Students are more likely to have broken a bone if they own a scooter: P = 0.5907.
The probability of having broken a bone in the group as a whole was only 0.5323.

ISBN: 9780170484084

Use the data in the following tables to answer the questions. Where needed, round your answers to 2 dp.

1 Data was collected from the same school about how many Year 9 and 10 students went to school camp. Complete the table.

	Year 9	Year 10	Totals
Went to camp	51	67	
Didn't go to camp	23	9	
Totals			

a What is the probability that a student went to camp? ______

b What is the probability that a randomly selected student was in Year 10 and went to camp?

______

c What is the probability that a Year 9 student went to camp?

______

d If two students were selected at random, what is the probability that both went to camp?

______

2 Maia surveyed 94 students. She asked them whether or not they play a musical instrument and whether they are left handed or right handed. Complete the table.

	Plays a musical instrument	Doesn't play a musical instrument	Totals
Left handed	6		**11**
Right handed		29	
Totals	**60**		

a What percentage of students are left handed?

______

b What is the probability that a randomly selected student plays a musical instrument and is right handed?

______

c What is the probability that a randomly selected right-handed student does not play a musical instrument?

______

d What is the probability that two randomly selected students from her survey both play musical instruments?

______

ISBN: 9780170484084

3 Theo did a survey of Year 11 students. He asked them whether or not they had ever been skiing.

	Had been skiing	Had not been skiing	Totals
Boys	11		**39**
Girls	16		
Totals			**84**

a What is the probability that a randomly selected student had been skiing?

____

b What is the probability that a randomly selected student was a boy who had been skiing?

____

c What is the probability that a randomly selected girl had been skiing?

____

d What is the probability that two randomly selected students from his survey had both been skiing?

____

4 A recent survey of 1200 students (540 primary and 660 secondary) recorded whether they ate dinner at a table or not. The probabilities are shown in the table below.

	Primary	Secondary	Totals
Dinner at a table		0.30	
Dinner not at a table	0.18		**0.43**
Totals	**0.45**		

a How many of the students surveyed ate dinner at a table?

____

b How many primary students did not eat dinner at a table?

____

c What is the probability that a primary student did not eat dinner at a table?

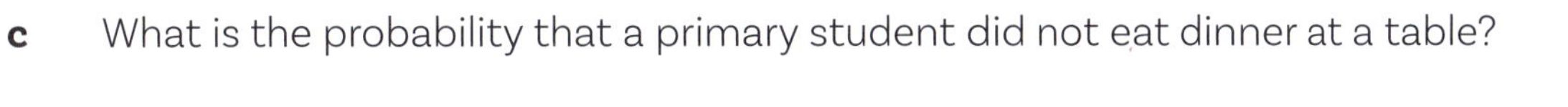

____

d How many students either were from a secondary school or ate dinner at a table?

____

ISBN: 9780170484084

Data analysis

There are two measures that we need to know in order to be able to discuss and compare distributions:

1 Where is the **centre** of the data?
2 How widely is the data **spread**?

Measures of centre

- There are **three** measures for the centre of the data.

Name	Calculation	Advantages	Disadvantages
Mean	$\frac{\text{the sum of all the data values}}{\text{the number of data values}}$	• Easy to calculate	• Distorted by values that are vastly different from the others.
Median	middle value	• Usually a good measure of centre	• Data must be put in order first
Mode	value that occurs most frequently	• Very easy to find	• Very unreliable as measure of centre • Often there are two or none

- Avoid using the word 'average' because it can have different meanings.

Mean

- The numbers **do not need to be in order** for this calculation.
- Means are not always whole numbers, so **sensible rounding** may be needed.
- The mean is influenced by unusually large or small values.

$$\textbf{mean} = \frac{\textbf{sum of all the data values}}{\textbf{number of data values}}$$

Example:

Notice that 0 must be included in the calculation.

0 2 9 4 7 6 8 7 53

$$\text{Mean} = \frac{0+2+9+4+7+6+8+7+53}{9}$$

$$= 10.\dot{6}$$

There are 9 numbers in the data set.

The mean of this data set is $10.\dot{6}$ or 10.7 (1 dp).

Notice what happens to the mean if the 53 is removed.

0 2 9 4 7 6 8 7

$$\text{Mean} = \frac{0+2+9+4+7+6+8+7}{8}$$

$$= 5.375$$

There are now 8 numbers in the data set.

The mean of this data set is 5.375.

The 53 was so much larger than the other values that it influenced the mean.

ISBN: 9780170484084

Median

- If there is an **odd number of values** in a data, the median is the **middle number**.
- If there is an **even number of values**, the median is **halfway between the two middle numbers** in the data set.
- Before you can calculate the median, you must **put the data in order**.

Examples:

1 A data set with an odd number of values

11 9 8 17 21 14 24 15 12

Put them **in order** before finding the median: **8 9 11 12 14 15 17 21 24**

This is the middle number.

The median of this data set = 14

2 A data set with an even number of values

0 2 3 5 7 11 16 17 18 19 19 22

The median for this data set = $\frac{11 + 16}{2} = 13.5$

Mode

- The mode is the **most common value**.
- Sometimes there are **several modes**.
- If there are **three or more** numbers that occur equally often, we say the data is **polymodal**.

Examples:

1 (8) 14 3 7 15 (8) 13 9 14 11 (8) 10 2

The most common number is **8**: there are three of them.

The mode of this data set is 8

2 11 9 (12) 17 2 (15) 6 10 (12) (15) 7 14 3

Both **12** and **15** occur twice.

The modes are 12 and 15

3 (21) 24 (27) (31) 26 (27) 30 22 (31) (21) 28

There are three numbers that occur equally often: **21**, **27** and **31**.

There are **three or more** modes, so the data is **polymodal**.

Calculate the mean, median and mode for these data sets. Round any calculations sensibly.

1 **1 1 1 2 2 3 3 4 4 5**

Mean = ____________ Median = ____________ Mode = ____________

2 **2 14 16 17 18 19 20 20 21 23 25**

Mean = ____________ Median = ____________ Mode = ____________

3 **22.1 23.2 25.6 27.4 28.3 29.0 32.8 34.1 40.5 43.5**

Mean = ____________ Median = ____________ Mode = ____________

4 **a** **1 2 2 3 5 7 8 9 11 12 89**

Mean = ____________ Median = ____________ Mode = ____________

b The mean and median are similar/different. Explain why.

____

5 **a** **2.1 2.9 3.4 3.7 3.8 4.1 4.2 4.8 5.0 5.2 5.6 5.9**

Mean = ____________ Median = ____________ Mode = ____________

b The mean and median are similar/different. Explain why.

____

6 The following sets of numbers are in ascending order. Find the missing numbers.

a 3 8 9 13 ☐ 19 20 24 – the median is 15.5.

b 3 4 5 7 10 11 ☐ 16 20 23 – the mean is 11.4.

7 Here are three data sets. Find the missing numbers.

Set A	2	18	6
Set B	13		
Set C	16		

Clues:

1 They all have the same mean.

2 Set C has the same median as set A.

3 The median of set B is double the median of set C.

 ISBN: 9780170484084

Measures of spread

Range

- The range is the **maximum** value **minus** the **minimum** value in the data set.
- Note: the range is a **single number**.
- Like the mean, the range is affected by unusually large or small values.
- The data does not need to be in order to calculate the range.
- The range is a measure of the **variability** of the data.

range = maximum – minimum

Quartiles

- The upper quartile (UQ) is the **median** (middle) of the **top half** of the data.
- The lower quartile (LQ) is the **median** (middle) of the **bottom half** of the data.
- The interquartile range (IQR) is also a measure of the **variability** of the data. It is usually considered to be a better measure of spread than range because it isn't affected by extreme values.

interquartile range (IQR) = upper quartile (UQ) – lower quartile (LQ)

Examples:

1 If the number of pieces of data is **odd**, **exclude** the median from the top and bottom halves of the data.

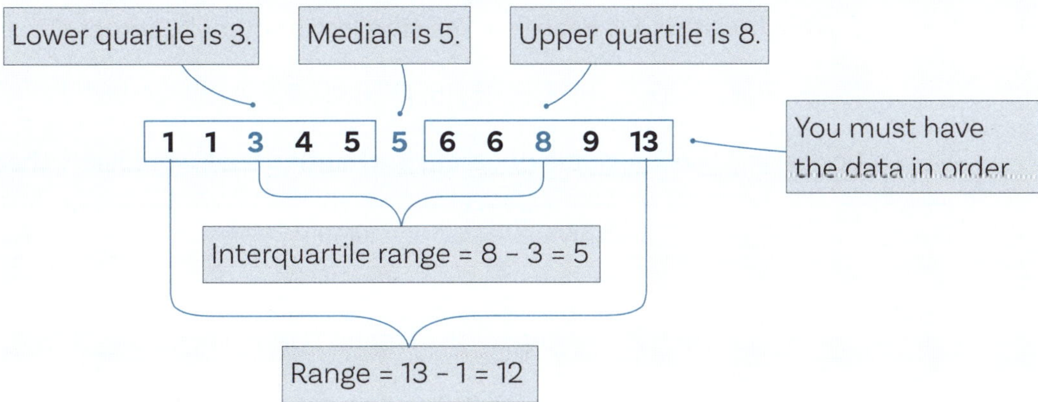

2 If the number of pieces of data is **even**, find the median values of the top and bottom halves of the data.

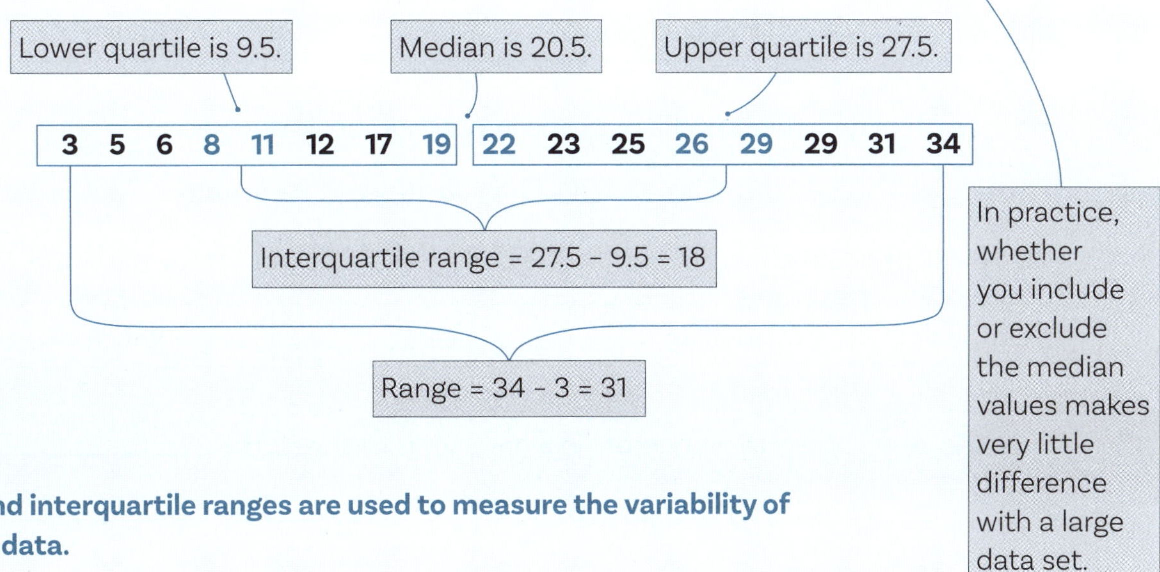

Ranges and interquartile ranges are used to measure the variability of groups of data.

Calculate the statistics for the following data sets.

1 **5 8 10 11 14 15 15 17 18 19 20 21 22 24 25**

Minimum = ______ LQ = ______ Median = ______ UQ = ______ Maximum = ______

Range = ________________ Interquartile range = ________________

2 **0 0 0 1 1 1 2 2 2 3 3 4 5 5 5 6 8 8**

Minimum = ______ LQ = ______ Median = ______ UQ = ______ Maximum = ______

Range = ________________ Interquartile range = ________________

3 **7 8 9 10 10 11 12 12 13 14 15 15 16 19 20 21**

Minimum = ______ LQ = ______ Median = ______ UQ = ______ Maximum = ______

Range = ________________ Interquartile range = ________________

4 The data set in question **1** is changed: the smallest number becomes 0 instead of 5.

0, 8, 10, 11, 14, 15, 15, 17, 18, 19, 20, 21, 22, 24, 25

Place a tick beside the measures of centre and spread that would stay the same.

Mean		Median		IQR		Range	

5 The data set in question **2** is changed: the largest number is changed to 20.

0, 0, 0, 1, 1, 1, 2, 2, 2, 3, 3, 4, 5, 5, 5, 6, 8, **20**

Place a tick beside the measures of centre and spread that would stay the same.

Mean		Median		IQR		Range	

6 The data set in question **3** is changed: the largest number is doubled.

7, 8, 9, 10, 10, 11, 12, 12, 13, 14, 15, 15, 16, 19, 20, **42**

Place a tick beside the measures of centre and spread that would stay the same.

Mean		Median		IQR		Range	

7 When the largest or smallest number in a data set is changed, the ________________ and the ________________ stay the same, but the ________________ and the ________________ become different.

ISBN: 9780170484084

PRACTICE SETS

Set 1

1 Simplify this expression:
$5x^2yz - 3x^2y + x^2y - xy + 2x^2yz - yx$

2 Write down the gradient of this line:

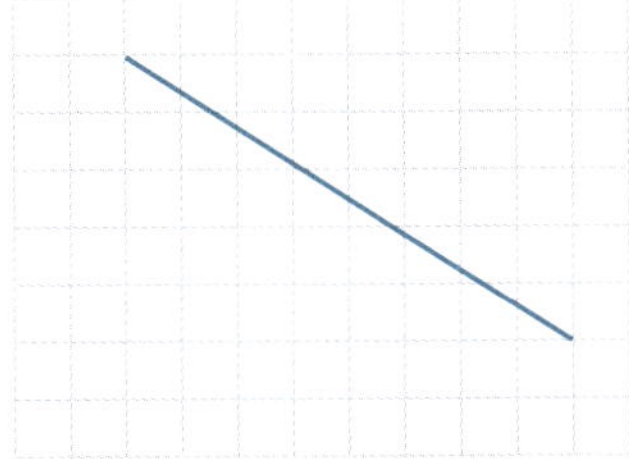

3 Convert 0.027 L into mL.

4 Expand and simplify this expression:
$(4 - 3x)(5x - 1)$

5 Calculate the size of angle x to 1 dp.

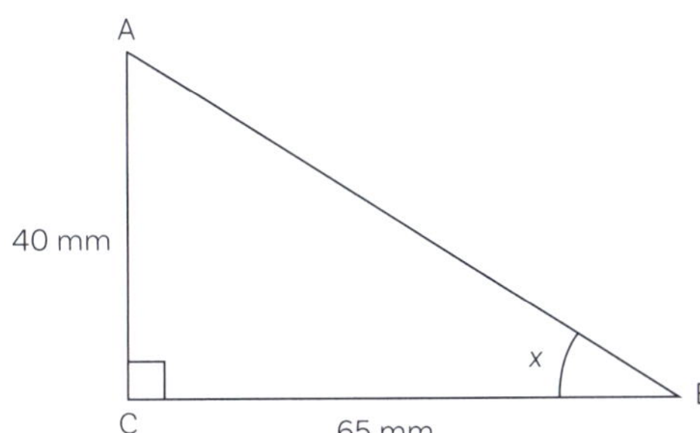

6 Calculate the area of this figure:

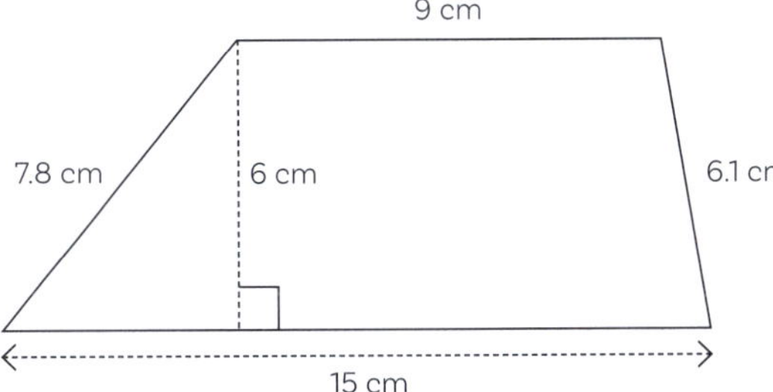

ISBN: 9780170484084

7 Calculate the value of this expression:

$32^{-\frac{1}{5}}$

8 Calculate the size of angle x.

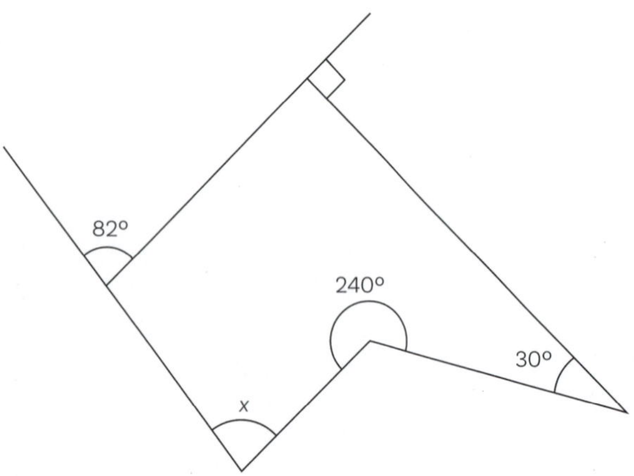

9 Fully factorise this expression:

$12x^2y - 3x^5 + 9x$

10 **a** Year 9 and 10 students were given a choice of Drama or Kapa haka. Complete this probability table:

	Year 9	**Year 10**	**Total**
Drama	24		**51**
Kapa haka		48	
Total	**60**		**135**

b Calculate the probability that a Year 9 student choose Drama.

c Calculate the probability that a student chose Kapa haka.

 ISBN: 9780170484084

Set 2

1 Factorise this expression:

$x^2 - 11x + 10$

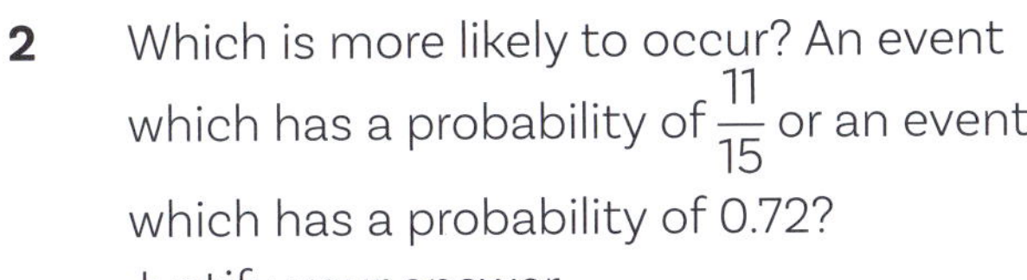

2 Which is more likely to occur? An event which has a probability of $\frac{11}{15}$ or an event which has a probability of 0.72? Justify your answer.

3 Calculate the length AC.

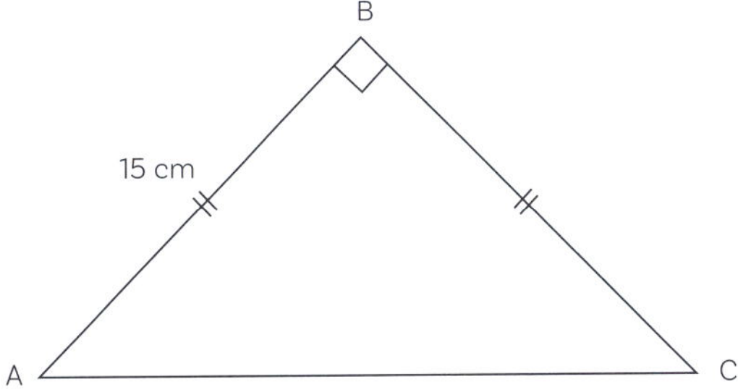

4 Simplify this expression:

$$\frac{-12x^2y^3z}{3x^5y}$$

5 Decrease 112 kg by 9%.

6 Write the equation of this line:

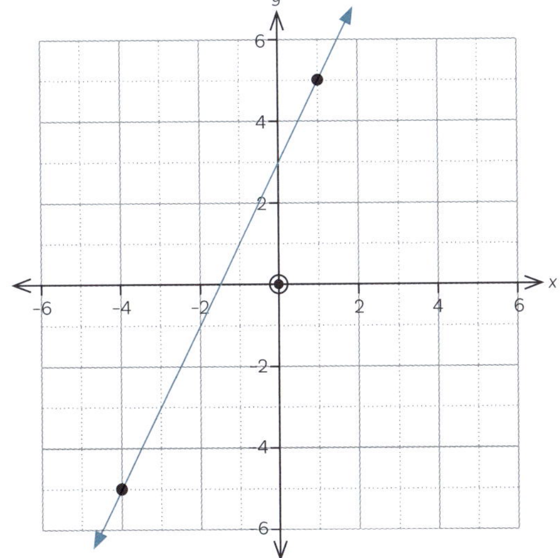

ISBN: 9780170484084

7 Calculate the volume of this cylinder. Round your answer to 4 sf.

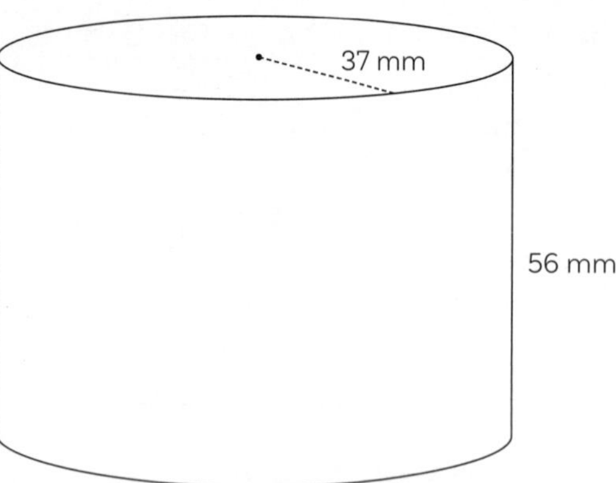

8 Simplify this expression:

$$\frac{4x^2}{3y} \div \frac{x}{y^2}$$

9 Calculate the size of an external angle of a regular heptagon. Round your answer to 1 dp.

10 Solve this equation:

$4(x - 5) = 23 - 6x$

11 A kitchen bench is 58 cm deep and 165 cm long. Calculate the area of the top in square metres.

12 Add these fractions:

$$\frac{5}{x-1} + \frac{2}{3x+2}$$

ISBN: 9780170484084

Set 3

1 The table represents points on a graph.
Write the equation for the line.

x	y
1	7
2	11
3	15
4	19
5	23

2 Simplify this expression:

$$\frac{3x + 5}{4} + x$$

3 Calculate the perimeter of this figure:

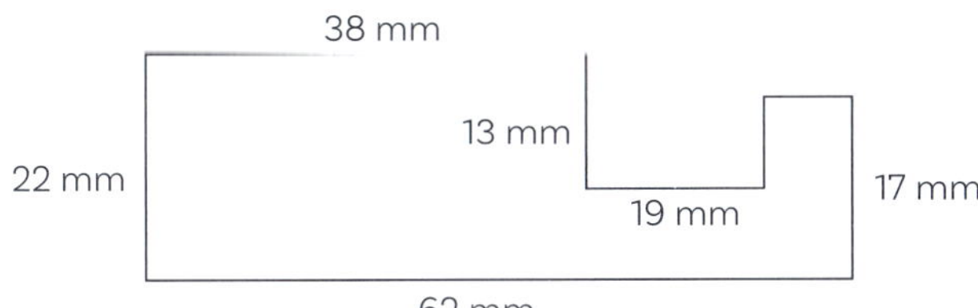

4 Solve this inequation:

$5(2x + 3) \leq 6(4 - x)$

5 The diagram shows a rectangle with a diagonal which is 76.32 cm long.
Calculate its height.

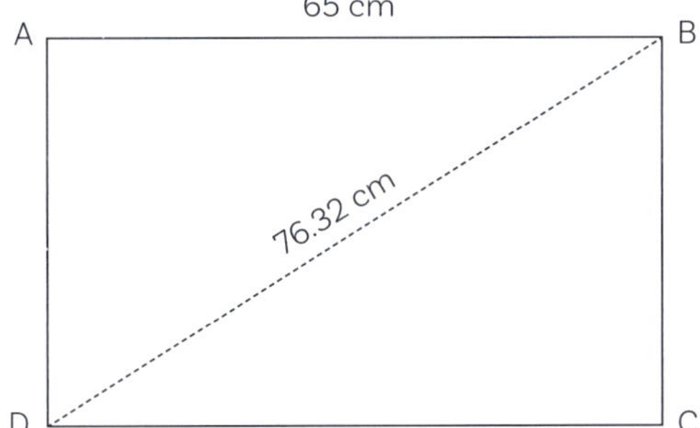

ISBN: 9780170484084

6 Simplify this expression:
$-2(-3x^3y)^2$

7 A recipe for muffins uses 1 cup of flour and $1\frac{1}{2}$ cups of cheese. Albert has only $\frac{3}{4}$ of a cup of flour. How much cheese should he use?

8 Calculate the size of angle x.

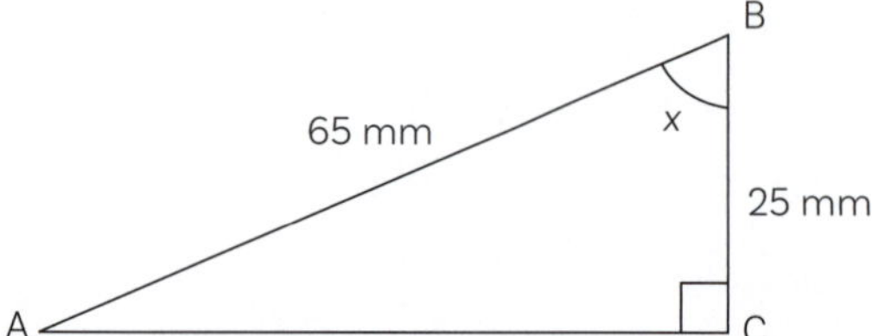

9 The mass of Adam's puppy increased from 4.50 kg to 5.58 kg. Calculate its percentage increase in mass.

10 **a** A class is at camp. Students could choose between rafting and kayaking in the morning, and the flying fox and abseiling in the afternoon. Complete the tree diagram.

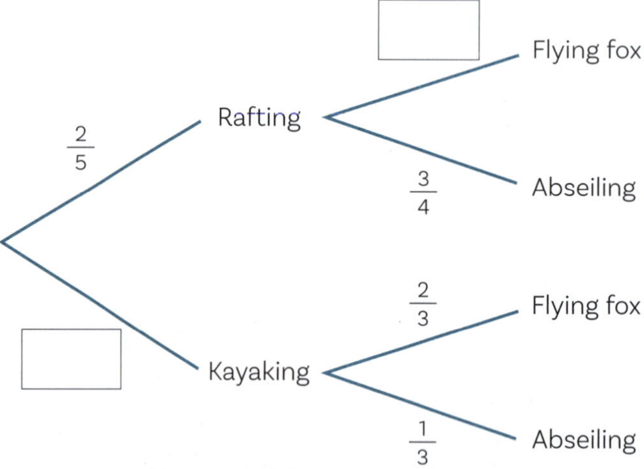

b Use the tree diagram to calculate the probability that a student wanted to go rafting and then abseiling.

c Use the tree diagram to calculate the probability that a student wanted to abseil.

ISBN: 9780170484084

Set 4

1 Draw the graph of $y = -\frac{1}{2}x + 1$.

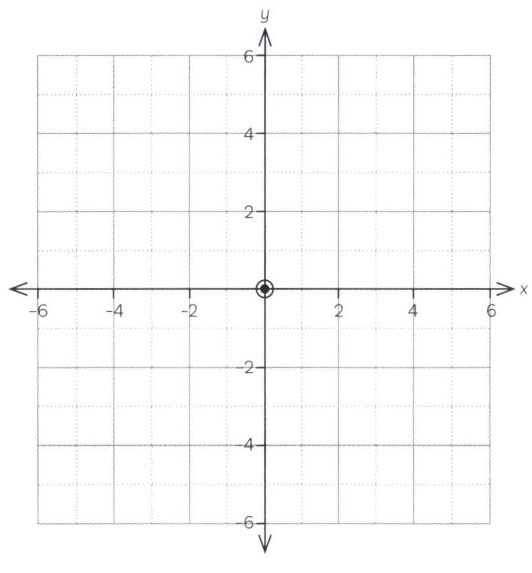

2 Simplify this expression:

$\sqrt{16x^{16}y^{36}}$

3 Calculate the area of a semicircle which has a radius of 15 cm.

4 Factorise this expression:

$x^2 - 5x - 14$

5 Calculate the length of AC.

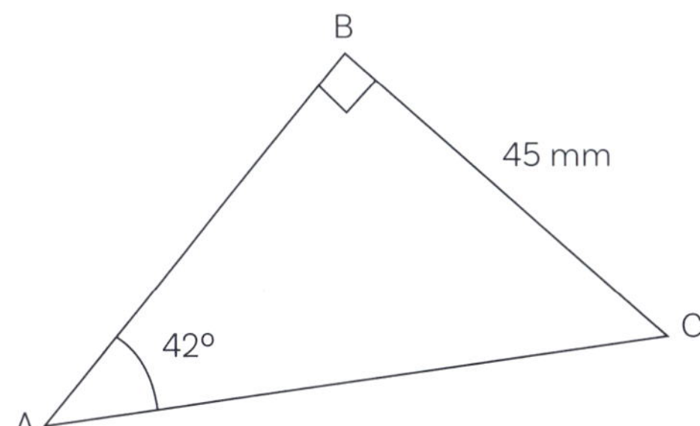

6 If $r = 3$, find the value of this expression. You may leave π in your answer.

$V = r^3 - \frac{1}{3}\pi r^3$

ISBN: 9780170484084

7 The ratio of the width of a TV screen to its height is 16:9. If a TV screen is 128 cm wide, how high is it?

8 Calculate the size of angle x.
Explain your reasoning.

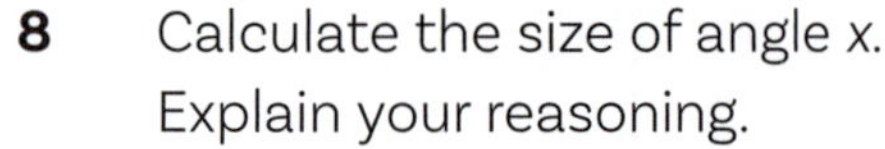

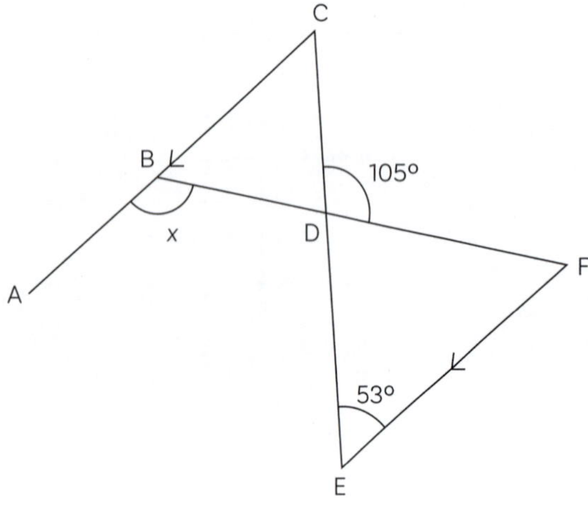

9 Solve this equation:

$$\frac{x}{2} + 5 - x = 7$$

10 Write 24 as a product of prime factors.

11 Write down the bearing shown by the blue arrow.

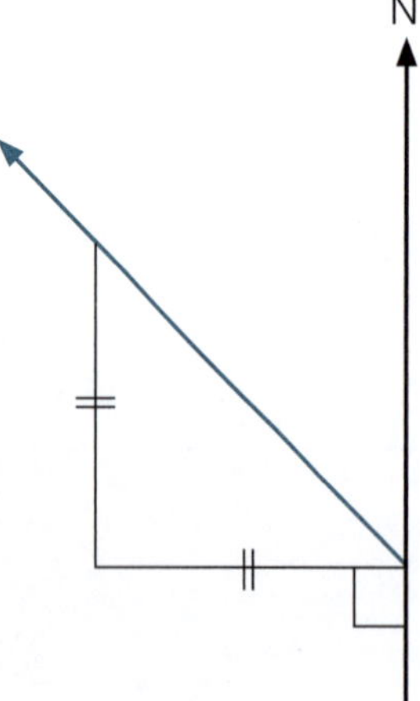

12 Increase \$68 by 14%.

ISBN: 9780170484084

Set 5

1 Rewrite the formula $A = \frac{(a + b)}{2}h$ with a as the subject.

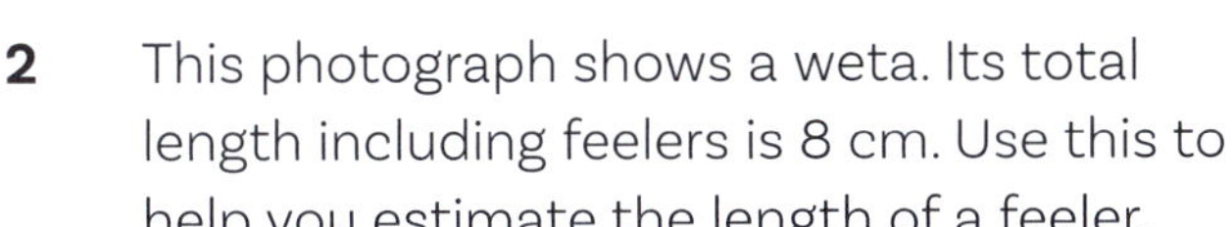

2 This photograph shows a weta. Its total length including feelers is 8 cm. Use this to help you estimate the length of a feeler.

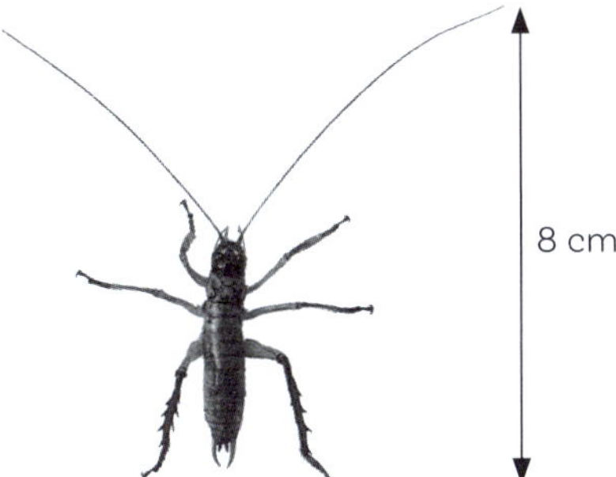

3 The probability that a person is colour blind in New Zealand is 0.045. How many colour-blind students would you expect there to be in a co-ed school with 1200 students?

4 Calculate the length of BC.

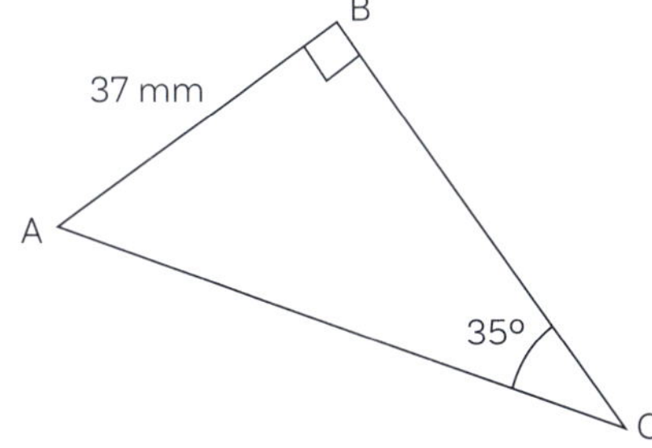

5 Simplify this expression:
$3(2x - 5) - x(4 - x)$

6 Solve this equation:
$2x^3 = -250$

7 Round 7109.95 to 4 sf.

8 The diagram shows a cuboid. Calculate the length of the line HB.

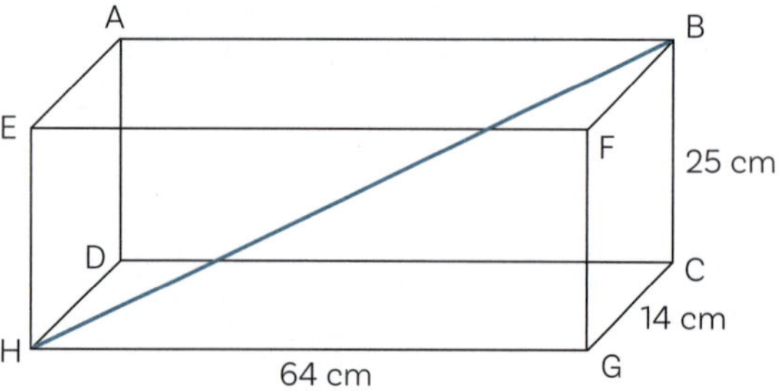

9 The retail price of a jacket is $112.70. Calculate its price without GST (15%).

10 Solve this equation:
$7 - 2x = 7x + 74$

11 Write the equation for a line which passes through the points (1, 3) and (5, –5).

12 Calculate the total surface area of this figure:

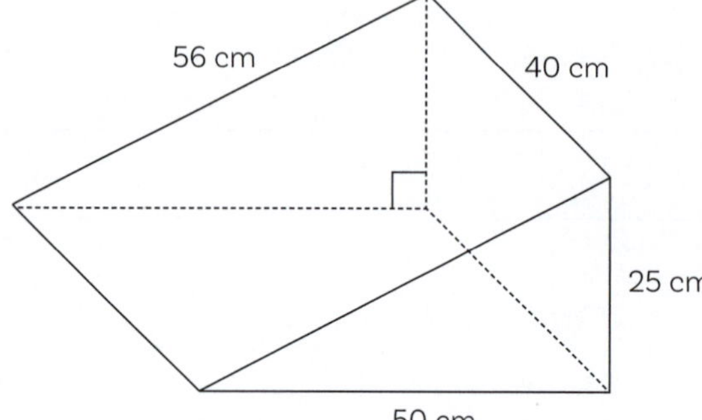

 ISBN: 9780170484084

ANSWERS

NUMBER (pp. 5–36)

Factors, multiples and primes (pp. 5–8)

Factors (p. 5)

1 1, 2, 4
2 1, 2, 3, 4, 6, 12
3 1, 3, 5, 15
4 1, 2, 3, 4, 6, 9, 12, 18, 36
5 1, 7
6 1
7 8: **1**, **2**, **4**, 8
12: **1**, **2**, 3, **4**, 6, 12
Common factors: **1**, **2**, **4**
8 5: 1, **5**
20: 1, 2, 4, **5**, 10, 20
Common factors: **1**, **5**
9 18: 1, 2, 3, **6**, 9, 18
60: 1, 2, 3, 4, 5, **6**, 10, 12, 15, 20, 30, 60
HCF: **6**
10 11: **1**, 11
21: **1**, 3, 7, 21
HCF: **1**
11 16: 1, **2**, 4, 8, 16
30: 1, **2**, 3, 5, 6, 10, 15, 30
HCF: **2**
12 17: **1**, 17
3: **1**, 3
HCF: **1**
13 **a** He can make 12 bags, which each contains 2 mint lollies and 3 fruit lollies.
b He can make 6 bags, which each contains 4 mint lollies, 6 fruit lollies and 5 caramel lollies.

Multiples (p. 6)

1 2, 4, 6, 8, 10, 12
2 5, 10, 15, 20, 25, 30
3 9, 18, 27, 36, 45, 54
4 12, 24, 36, 48, 60, 72
5 10, 20, 30, 40, 50, 60
6 1, 2, 3, 4, 5, 6
7 2: 2, 4, **6**, 8, 10, 12
3: 3, **6**, 9, 12, 15, 18
LCM: **6**
8 5: 5, 10, **15**, 20, 25, 30
3: 3, 6, 9, 12, **15**, 18
LCM: **15**
9 6: 6, 12, **18**, 24, 30, 36
9: 9, **18**, 27, 36, 45, 54
LCM: **18**
10 8: 8, 16, **24**, 32, 40, 48
12: 12, **24**, 36, 48, 60, 72
LCM: **24**
11 10: 10, 20, 30, 40, 50, **60**
12: 12, 24, 36, 48, **60**, 72
LCM: 60
12 7: 7, 14, 21, 28, 35, 42, 49, 56, **63**
9: 9, 18, 27, 36, 45, 54, **63**
LCM: **63**
13 **a** The 40th bead. **b** The 120th bead.

Prime numbers (p. 7)

1

1	**2**	**3**	4	**5**	6	**7**	8	9	10
11	12	**13**	14	15	16	**17**	18	**19**	20
21		**23**	24	25	26	27	28	**29**	30
31	32	33	34	35	36	**37**	38	39	40
41	42	**43**	44	45	46	**47**	48	49	50

2 No because 2 is prime.
3 No because 1 is not prime.
4 Yes, 53
5 Odd Even

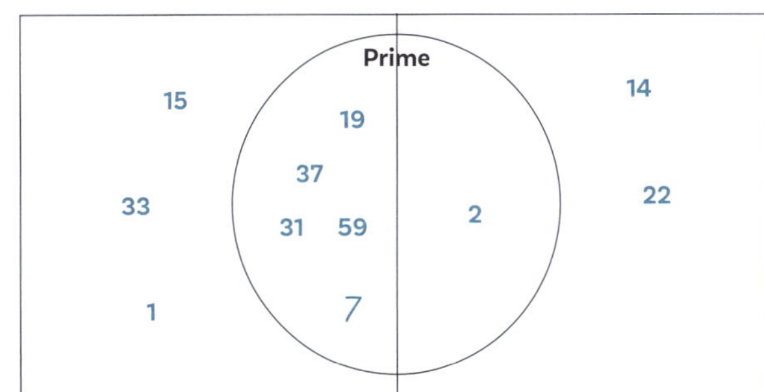

Prime factorisation (p. 8)

1

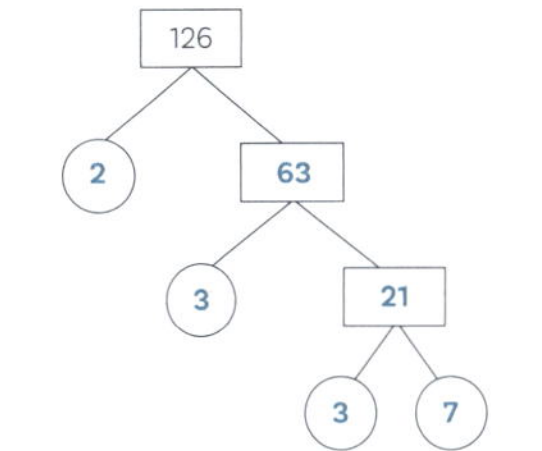

126 = **2 x 3² x 7**

2

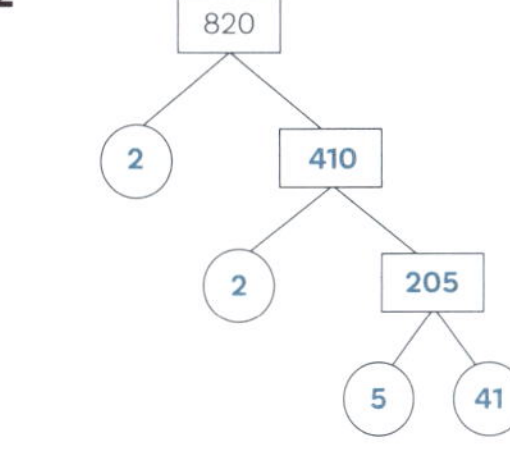

820 = **2² x 5 x 41**

3

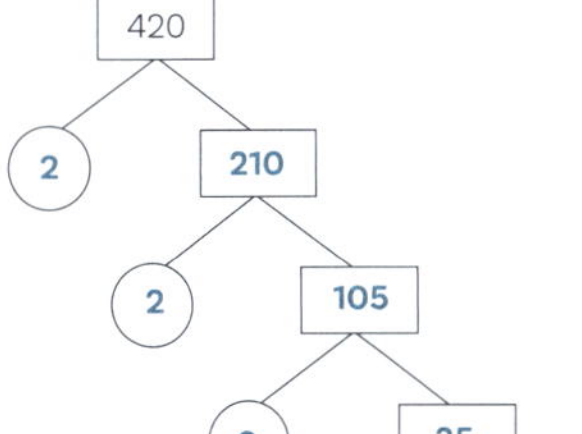

420 = **2² x 3 x 5 x 7**

4

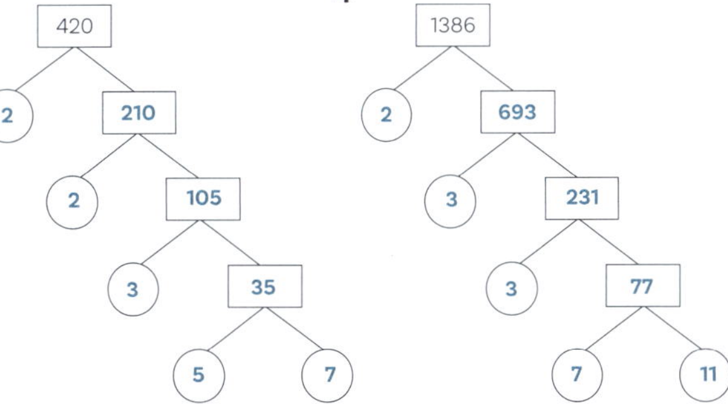

1386 = **2 x 3³ x 7 x 11**

ISBN: 9780170484084

Powers and roots (pp. 9–11)

Powers (p. 9)

1 36 **2** 256
3 -8 **4** 81
5 -9 **6** 125
7 $\frac{1}{4}$ **8** 0.0625
9 5^2 **10** 3^3
11 -2^5 or $(-2)^5$ **12** 10^3
13 10^6 **14** 0.1^2

Fractional powers — roots (p. 10)

1 4 **2** 5
3 5 **4** 2
5 0.5 **6** $\frac{1}{2}$

Square root	Lower/upper limits	Answer
$\sqrt{40}$	6 and 7	6.325
$\sqrt{6}$	2 and 3	2.449
$\sqrt{80}$	8 and 9	8.944
$\sqrt{67}$	8 and 9	8.185
$\sqrt{20}$	4 and 5	4.472

Negative powers (p. 11)

1 $\frac{1}{2}$ **2** $\frac{1}{100} = 0.01$
3 $\frac{1}{16}$ **4** $\frac{1}{1\,000\,000} = 0.000001$
5 $\frac{4}{5}$ **6** 3
7 $\frac{1}{4}$ **8** 3

Mixing it up (p. 11)

1 18 **2** 192
3 -16 **4** -16
5 16 **6** 12
7 $-\frac{1}{25}$ **8** 4
9 9 **10** 145
11 48 **12** 72

Rounding (pp. 12–16)

1 Rounding to decimal places (pp. 12–13)

1 6.82 **2** 2.47
3 64.1 **4** 142.1
5 9.091 **6** 91.446
7 399.90 **8** 0.03
9 0.00 **10** 1.02
11 3.0 **12** 5.6
13 9.43 **14** 2028
15 8.3 **16** 2.851
17 **a** 38°C **b** 38.2°C
18 **a** 5094 m² **b** 5090 m²

2 Rounding to significant figures (pp. 14–16)

Number	Number of significant figures
51.2	3
13 706	5
14.028	5
94 778.5	6
10	1
3.20046	6
4 461 030	6
12 000	2
1.20	3
1205	4
0.2	1
0.004	1
43.50	4
9.670420	7
53.060	5
2.0670	5
79 000 001	8
10 000	1
60.0850	6
0.060410	5
1.030	4
0.090	2

1 53 **2** 35 000
3 0.068 **4** 10 000
5 110 **6** 0.00106
7 742 000 **8** 0.00694
9 1 672 000 **10** 0.07000
11 70 **12** 90 000
13 0.5 **14** 0.06
15 0.1 **16** 0.008
17 0.942 **18** 0.108
19 2.17 **20** 1.79
21 Lowest: 10.50 + 11.50 = 22.00
Highest: $11.4\dot{9} + 12.4\dot{9} = 23.\dot{9}$ ∴ less than 24

Rates (pp. 17–18)

1 $13.75 **2** $76.65
3 24 min **4** 84 kph
5 3600 kg **6** 37.5 sec
7 $0.81\dot{6}$m **8** 9000 cans

Exchange rates (p. 19)

1 €1625 **2** $US1850
3 £1225 **4** $NZ230.77
5 $NZ810.81 **6** $NZ30.61

Ratios (pp. 20–24)

Simplifying ratios (pp. 20–21)

1 3:2 **2** 10:7:3
3 4:5 **4** 2:3
5 7:5 **6** 6:4:3
7 1:5 **8** 1:3:2
9 2:3:6 **10** 24:1
11 6:1 **12** 4:7:1

ISBN: 9780170484084

Ratio calculations where the total amount is given (pp. 21–22)

1 Sam 34 lollies : Sarah 17 lollies
2 1 cup liquid : $1\frac{1}{2}$ cups dry ingredients
3 3L cordial : 21 L water
4 39 trillion bacteria cells : 30 trillion human cells
5 632 people
6 540 000 tonnes butter : 324 000 tonnes cheese
7 Cat $15 200, older nieces $3800, youngest niece $7600
8 428 m^2
9 Open space: 14 ha
High-density housing: 21 ha
Medium-density housing: 15.75 ha
Low-density housing: 5.25 ha

Ratio calculations where one part is given (pp. 23–24)

1 25 615 000 sheep
2 1.4 mm
3 350 g dried fruit
4 28 students
5 600 muscles
6 481 entries
7 $60 per week
8 164 000 species
9 398 tosses
10 2450 m
11 6.25 cm

Scale diagrams (pp. 25-26)

1 $\frac{6}{90} = \frac{5}{S}$
$S = 75$ m
2 $\frac{2.1}{48} = \frac{W}{14}$
$W = 0.6125$ m
3 $\frac{225}{50} = \frac{W}{12}$
$W = 54$ mm
4 $\frac{10}{60} = \frac{L}{25}$
$L = 4.1\dot{6}$ m

Percentages (pp. 27–29)

Using percentages (pp. 27–28)

1 76.85 km
2 30.6 g
3 1.56 kg
4 $907.20
5 5.04 mL
6 $175
7 942.08 m
8 $68.25
9 70.08 g
10 3.2 min
11 72.24 cm
12 4.4 cm
13 45%
14 78 students
15 $389.35
16 16.67% (2 dp)
17 Private cars 57.89%, motor bikes 23.68%, taxis 10.53%, delivery vans 7.89%
18 Nieces get $4560 each, cat gets $15 200
19 240 000 plastic fragments

Calculating percentage changes (p. 29)

1 Increase of 5%
2 Decrease of 10%
3 Increase of 15%
4 Decrease of 25%
5 Increase of 2%
6 Decrease of 1%
7 Increase of 125%
8 Decrease of 86%
9 She lost 24%
10 It increased by 12.47%

GST (pp. 30–31)

Adding GST (p. 30)

1 $98.90
2 $17.83
3 $103.44
4 $117.67
5 $4665.54
6 $65.26
7 $0.26
8 $74.98
9 $609.50
10 **a** $4025.00 **b** $525.00
11 **a** $3.99 **b** $0.52

Calculating the pre-GST price and GST (p. 31)

1 $85.22
2 $23.87
3 $95.64
4 $223.29
5 $0.86
6 $852.74
7 $1.38
8 $75.39
9 $773.91
10 $652.04
11 $0.45
12 The salesman deducted 15% of the full price ($5999), which is $899.85, resulting in a sale price of $5099.15. He should have deducted 15% of the GST-exclusive price ($5216.52), which is $782.48, producing a sale price of $5216.52. So he sold the car too cheaply.

Interest (pp. 32–33)

Simple interest (SI) (p. 32)

1 $525
2 $24 750
3 $1400
4 5.5%
5 6.25%
6 4 years

Compound interest (CI) (p. 33)

1 $8427.00
2 $30 198.74
3 $100 943.01
4 **a** $4416.96 **b** $396.28 more

Standard form (pp. 34–36)

Power of ten	Fraction	Whole number or decimal
10^3		**1000**
$\mathbf{10^{-1}}$	$\frac{1}{10}$	**0.1**
$\mathbf{10^6}$		1 000 000
10^{-2}	$\mathbf{\frac{1}{100}}$	**0.01**
$\mathbf{10^{-4}}$	$\mathbf{\frac{1}{10\,000}}$	0.0001
$\mathbf{10^{-8}}$	$\frac{1}{100\,000\,000}$	**0.00000001**
10^0		**1**

Standard form to ordinary numbers (pp. 34–35)

1 940
2 142 000
3 2690
4 6 111 000
5 8.0
6 0.045
7 0.9467
8 0.000037
9 0.0062
10 0.0000000183

Ordinary numbers to standard form (pp. 35–36)

1. 5.43×10^2
2. 1.2×10^3
3. 7.42×10^1
4. 1.689×10^0
5. 7.673×10^6
6. 8.005×10^4
7. 3.666×10^8
8. 1×10^0
9. 5.1×10^{-1}
10. 6.7×10^{-4}
11. 1.4×10^{-2}
12. 7×10^{-6}
13. 1×10^{-4}
14. 8.32×10^{-3}
15. 1.101×10^{-3}
16. 4×10^{-10}

ALGEBRA (pp. 37–95)

Simplifying expressions (pp. 37–38)

Multiplying and dividing (p. 37)

1. y^3
2. adf
3. $7d$
4. $6df$
5. $-3efg$
6. $7df^2$
7. $-10a^2$
8. $-24abc$
9. $-5cd^2$
10. $18f$
11. $\frac{g}{2}$
12. $\frac{14d}{e}$
13. $\frac{3}{e}$
14. $\frac{20dg}{f}$
15. $\frac{12df}{ac}$
16. $\frac{12efg}{a^3}$

Adding and subtracting (p. 38)

1. like
2. unlike
3. unlike
4. like
5. like
6. like
7. like
8. unlike
9. unlike
10. like
11. $4x$
12. $6x^2 + x$
13. $5 + 2ab$
14. $9p^5q + 3pq^5$
15. $8y^2 - y^3$
16. $x^2 + 12x + 35$
17. $11x - 4y + 1$
18. $6ab + 4a - 3b$
19. $17 + q - 6p$
20. $x^2 - 12x + 27$
21. $y^2 + 7y - 8$
22. $3x^2 + 6xy - 7y^2$
23. x^2
24. $5x^2yz + xyz^2$

Powers (pp. 39–41)

Multiplying terms that have powers (p. 39)

1. p^5
2. a^5
3. b^9
4. g^9
5. $6a^3$
6. $10a^3b^2$
7. $30b^{15}$
8. f^3g^7
9. $15y^{10}$
10. $-a^3b^4$
11. $-6y^6$
12. $8z^{10}$

Dividing terms that have powers (p. 40)

1. y^6
2. y^3
3. b^9
4. $5b^5$
5. $\frac{y^3}{2}$
6. x^2y
7. $4y^2z$
8. $-\frac{1}{2x}$
9. $\frac{3x}{y^5}$
10. $-\frac{x}{3}$
11. $\frac{1}{3x}$
12. $\frac{2y^2z^2}{x^3}$
13. $\frac{1}{6}$
14. $\frac{1}{2xz}$

Powers of powers (p. 41)

1. b^8
2. $16x^{12}$
3. a^{14}
4. $-y^{15}$
5. $9x^8$
6. $-64x^{12}$
7. $-50x^6$
8. $x^6y^{12}z^9$
9. $-9a^6b^4$
10. $18x^2y^6$
11. $\frac{x^6}{8}$
12. $\frac{x^8}{25}$

Roots (p. 42)

1. 5
2. x^5
3. 10
4. $6x^3$
5. $4x^8$
6. 4
7. $10x^{50}$
8. $7x^4y^3$
9. $9x^{64}$
10. $8x^6y^{12}$
11. $5x^5y^3$
12. $8x^{32}y^8z^{18}$

Expanding and factorising (pp. 43–45)

Expanding brackets (p. 43)

1. $15x - 20$
2. $12x - 6x^2$
3. $-6x^2 + 10x$
4. $21x - 16$
5. $6x - 12$
6. $16x - 21$
7. $2 - 6x$
8. $24x + 23$
9. $2x^2 - 8x + 4$
10. $23x - 24$
11. $14x^2 - 6x$
12. $-23x^2 - 27x$

Factorising (pp. 44–45)

1. $8(2x + 1)$
2. $12(3x - 5)$
3. $20(5 + x)$
4. $12x(5 - 2x)$
5. $4xy(3y + 1)$
6. $ab^3c^3(b + ac)$
7. $5(x - 3)$
8. $x(x - 3)$
9. $8(x + 2)$
10. $6(4 - 3a)$
11. $2(5y + 3)$
12. $x(x + 11)$
13. $x(2 + 3x)$
14. $5x(x + 3)$
15. $4y(y + 3)$
16. $2a(2a + 1)$
17. $3x(x + 5)$
18. $7y(3 - 2y^2)$
19. $2a^2(5a + 1)$
20. $4b^2(3 + 4b)$
21. $xy(1 + 9xy)$
22. $7xy(2x + y)$
23. $abc(2b - ac^2)$
24. $8abc^3(3b - a)$
25. $x(10y + 5x - y^2)$
26. $4xy(3y + x - 2xy)$

Algebraic fractions (pp. 46–51)

Simplifying fractions (p. 46)

1. $8a + 9b$
2. $\frac{1}{7}$
3. a
4. $7x + 5x^2$
5. $\frac{1}{2}$
6. $\frac{x - 1}{3}$
7. $\frac{y}{4 + 2x}$
8. $\frac{7a - 6}{7}$
9. $\frac{3b + 4a}{5}$
10. $3y^2 + 2xy$
11. $\frac{c}{7b - 5}$
12. $2 + a$

Multiplying fractions (p. 47)

1. $\frac{18x}{y^2}$
2. $\frac{14}{z^2}$
3. $\frac{8z}{35}$
4. $\frac{2y}{9}$
5. $\frac{5}{6z}$
6. $\frac{y}{6}$
7. $\frac{3(y - 5)}{20}$ or $\frac{3y - 15}{20}$
8. $\frac{y^2z^3}{7}$
9. $\frac{4z}{15}$
10. $\frac{3x(x + 1)}{10}$ or $\frac{3x^2 + 3x}{10}$
11. $\frac{2xy^2z^2}{5}$
12. $\frac{y}{2(y - 1)}$ or $\frac{y}{2y - 2}$

Dividing fractions (p. 48)

1. 6
2. $\frac{y}{x}$
3. $\frac{3}{2}$
4. $3y$
5. 25
6. $12x^2$
7. $\frac{1}{y^2}$
8. $\frac{5}{2}$
9. $\frac{12x}{5}$
10. $\frac{25x}{y}$

ISBN: 9780170484084

Adding and subtracting fractions (pp. 49–51)

1 Where the denominators are the same (p. 49)

1	$\frac{5x}{9}$	2	$\frac{9}{x}$	3	$\frac{x}{7}$
4	$\frac{6x}{y}$	5	$\frac{5x}{11}$	6	$\frac{x}{2}$
7	x	8	$\frac{6}{x}$	9	$\frac{6x-1}{7}$
10	$\frac{11x-1}{9}$	11	$\frac{x-2}{2}$	12	$1-x$

2 Where the denominators are different (p. 50)

1	$\frac{11x}{12}$	2	$\frac{13x}{14}$	3	$\frac{7x}{15}$
4	$\frac{x}{14}$	5	$\frac{5xy-3}{15}$	6	$\frac{24-x}{6}$
7	$\frac{3y-2x}{12}$	8	$\frac{17+x}{5}$	9	$\frac{7x-5}{5}$
10	$\frac{4x-5}{6}$				

3 Where the denominators are different algebraic expressions (p. 51)

1	$\frac{7x-5}{(x+1)(2x-4)}$	2	$\frac{-3x^2+5x+3}{(3x-2)(x+1)}$
3	$\frac{2x^2+8x+14}{(x+1)(x+5)}$	4	$\frac{7x^2-8x+10}{(3x-2)(x+4)}$

Substitution (p. 52)

1	-3	2	9	3	48
4	-5	5	-2	6	-1
7	31	8	2	9	37
10	36π	11	36π	12	$27-9\pi$

Solving linear equations (pp. 53–57)

One-step equations (p. 53)

1	1.7	2	3.1	3	2.8
4	2	5	-47	6	-0.68
7	11.5	8	0.6	9	0.1
10	-133				

Two-step equations (p. 54)

1	3.8	2	10	3	3
4	41.75	5	-20	6	-0.04
7	0.2	8	16	9	47
10	0.005				

Equations with a variable on both sides (p. 55)

1	10	2	-2	3	9
4	-5	5	20	6	3
7	5	8	-122	9	4
10	-0.1				

Equations with brackets (p. 56)

1	-22	2	-3	3	-1
4	1.5	5	1	6	1
7	3	8	2.2		

Equations with fractions (p. 57)

1	35	2	10	3	3.75
4	5	5	-0.6	6	0.25
7	1	8	7		

Equations with powers (p. 58)

1	±9	2	3	3	±2
4	-4	5	2	6	±7
7	-3	8	±2	9	±6
10	-4				

Inequations (p. 59)

1	$x > 8$	2	$x < 11$
3	$x \geq 4$	4	$x < -7$
5	$x > 1$	6	$x \geq -43$
7	$x > -5.5$	8	$x \leq 1$

Rearrangement of expressions (pp. 60–62)

1 Where the subject appears once (pp. 60–61)

1	$r=\sqrt{\frac{A}{\pi}}$	2	$P=\frac{100I}{RT}$
3	$x=\frac{y-c}{m}$	4	$x=\frac{y-8}{5}$
5	$x=\sqrt{\frac{y-b}{a}}$	6	$x=\frac{3y-12}{2}$
7	$x=\frac{y-25}{6}$	8	$r=\sqrt{\frac{3V}{h}}$
9	$x=\frac{6y-29}{2}$	10	$x=\frac{4y+30}{3}$
11	$a=b$	12	$h=\frac{2A}{a+b}$
13	$p=\frac{a-3tq}{3t}$	14	$C=\frac{5(F-32)}{9}$

2 Where the subject appears twice (p. 62)

1	$a=\frac{6}{2+b}$	2	$a=\frac{3b}{5+b}$
3	$a=\frac{2}{2-b}$	4	$a=\frac{7b}{6+5b}$
5	$a=\frac{4}{b+1}$	6	$a=\frac{3b}{2b+1}$

Quadratic expressions (pp. 63–66)

Expanding quadratic expressions (pp. 63–64)

1	$x^2+7x+12$	2	$x^2+5x-14$
3	$x^2-6x-27$	4	$x^2-13x+30$
5	$x^2+13x+40$	6	$3x^2-20x-7$
7	$x^2+16x+64$	8	$x^2-10x+25$
9	$3x^2-5x-2$		
10	$6x^2+16x-6$ or $2(3x^2+8x-3)$		
11	$9-6x+x^2$ or x^2-6x+9		
12	$9x^2-30x+25$	13	$6x^2-15x$
14	x^2-1	15	$20x^2+7x-6$
16	$28+13x-6x^2$	17	$16x^2-9$
18	$25-4x^2$		

Factorising quadratic expressions (pp. 65–66)

1	$(x+2)(x+4)$	2	$(x+7)(x+1)$
3	$(x+4)(x+9)$	4	$(x+6)(x-2)$
5	$(x+15)(x+4)$	6	$(x-6)(x-3)$
7	$(x+4)(x-6)$	8	$(x-5)^2$
9	$(x+1)(x-6)$	10	$(x+2)(x-3)$
11	$(x-2)(x-3)$	12	$(x+6)(x-1)$
13	$(x+5)(x-5)$	14	$(7+x)(7-x)$
15	$(3+4x)(3-4x)$	16	$(5+3x)(5-3x)$
17	$(3+x)(8+x)$ or $(x+3)(x+8)$		
18	$(3-x)(5-x)$		

Straight lines (pp. 67–95)

Coordinates (p. 67)

1	A	(2, 7)	B	(3, -2)
	C	(-2, 9)	D	(0, 2)
	E	(-7, -3)	F	(8, 0)

2

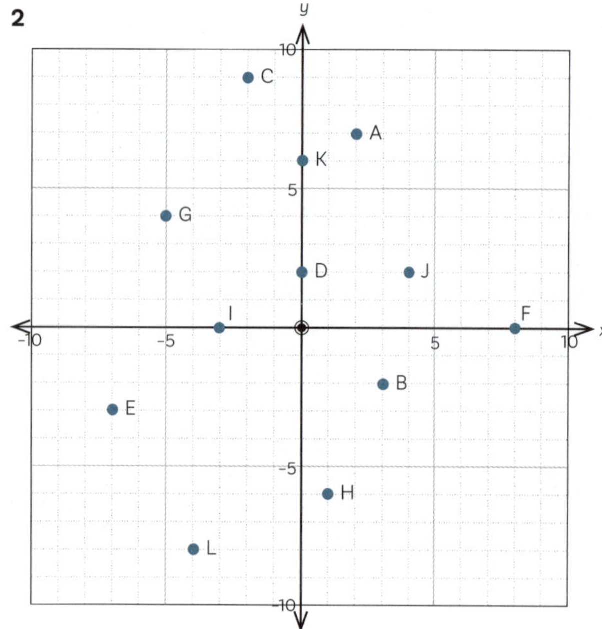

Linear patterns with discrete data (pp. 68–72)

1 **a** $B = 3n + 1$

b

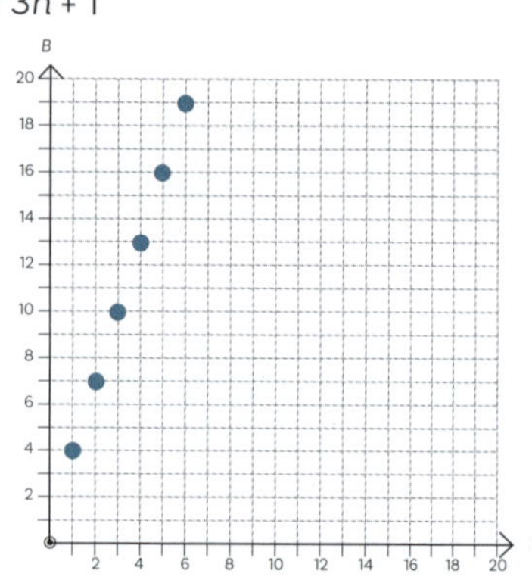

c 30th pattern needs 91 buttons.

d The 90th pattern would need 271 buttons.

e 3 represents how many buttons need to be added for each extra pattern.
1 represents the single button at the centre of every pattern.

2 **a** $B = 2n + 4$

b

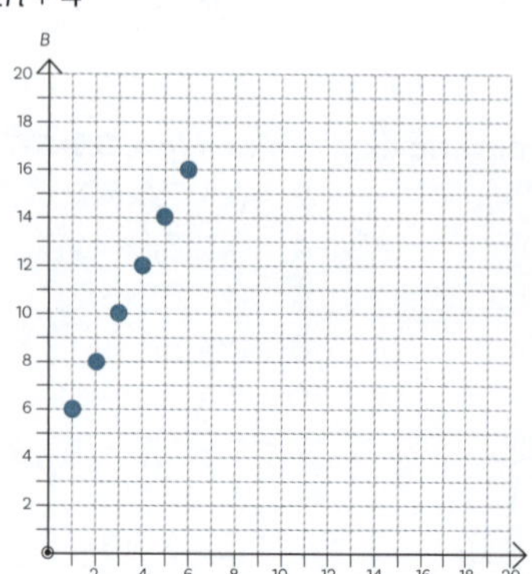

c 30th pattern needs 64 buttons.

d The 80th pattern would need 164 buttons.

e 2 represents the number of buttons that needs to be added for each extra pattern.
4 represents the 4 buttons at the centre of every pattern.

3 **a** $M = 4n + 1$

b

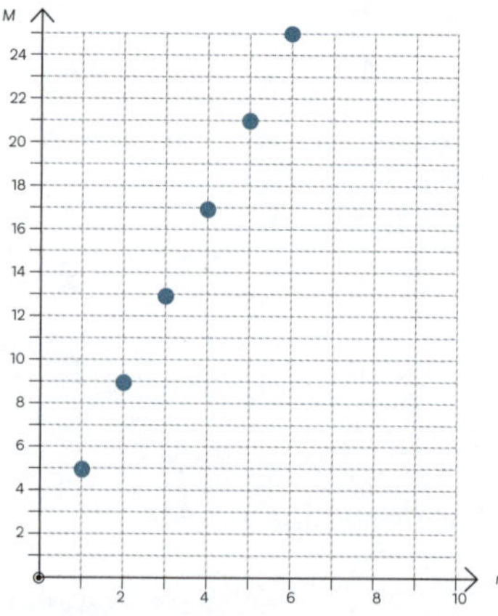

c 30th pattern needs 121 matchsticks.

d The 61st pattern would need 245 matchsticks.

e 4 represents the number of matchsticks that needs to be added for each extra pattern.
1 represents the single upright matchstick on the left.

4 **a** $D = 10s + 40$

b

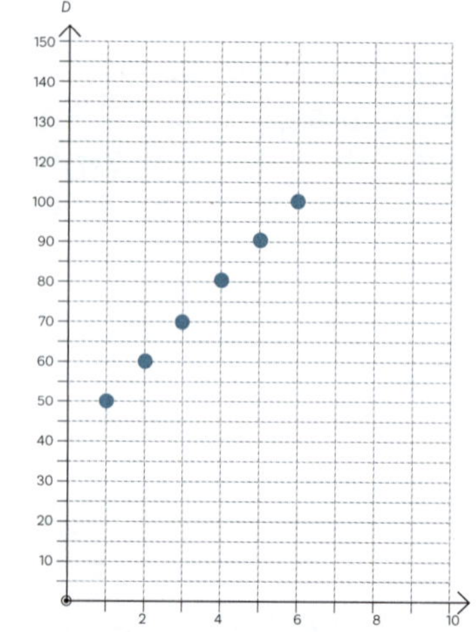

c If he passes 13 standards with Merit or Excellence, he earns $170.

d If he earned $210, he passed 17 standards with Merit or Excellence.

e 10 represents how much he earns for each standard he passed with Merit or Excellence.
40 represents the $40 he gets for passing Level 1.

The gradient of a line (pp. 73–75)

1 **a** $\frac{4}{2} = 2$ **b** $\frac{3}{2}$

c $\frac{2}{5}$ **d** $\frac{0}{5} = 0$

e $-\frac{2}{4} = -\frac{1}{2}$ **f** $-\frac{5}{2}$

g $\frac{2}{2} = 1$ **h** $\frac{3}{0}$ = undefined

i $\frac{1}{3}$ **j** $-\frac{1}{7}$

 ISBN: 9780170484084

2 Your answer might have a different length but the slope must be the same.

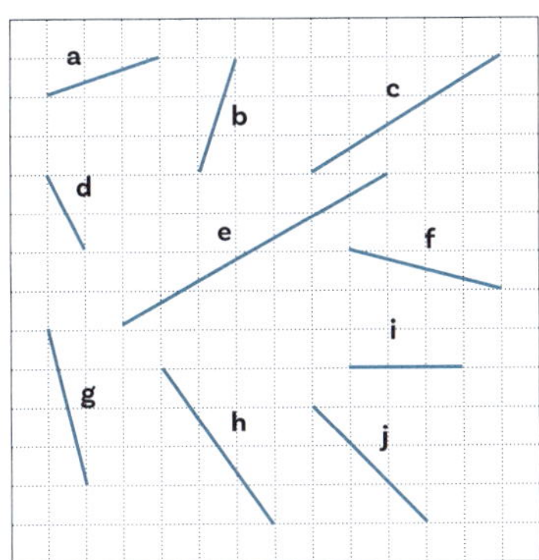

Gradients with different scales (pp. 76–79)

1 Drawing gradients (pp. 76–77)

1

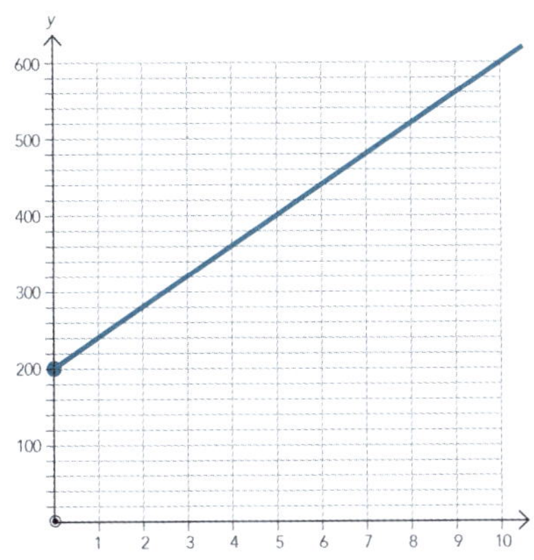

2

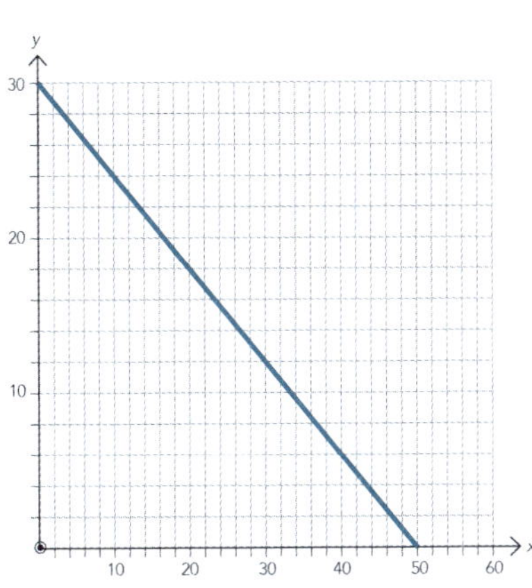

3

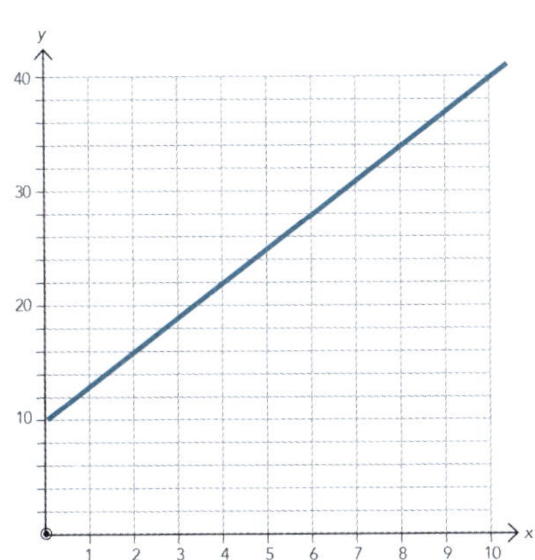

4

5

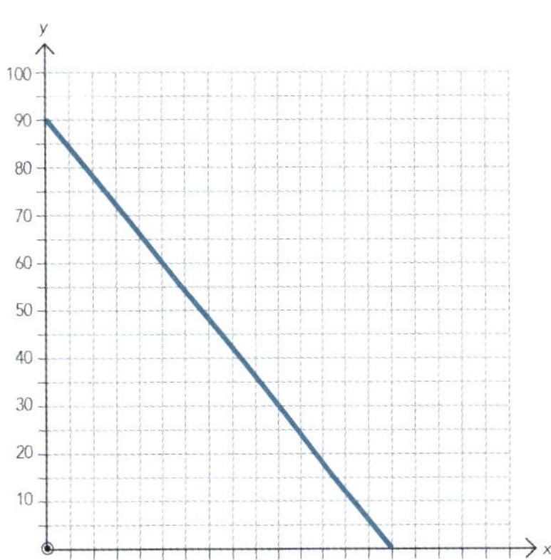

6

7

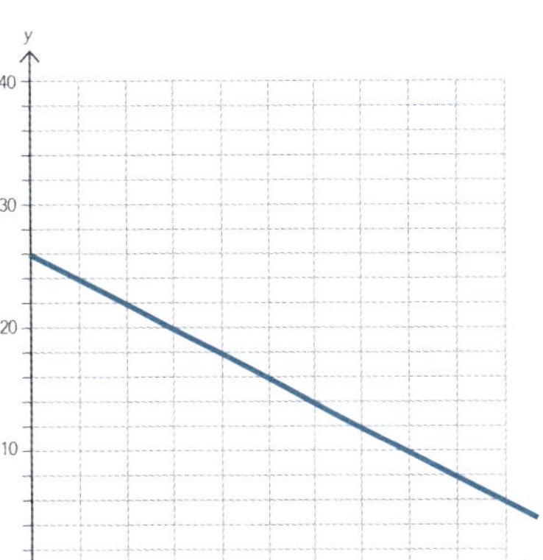

ISBN: 9780170484084

8

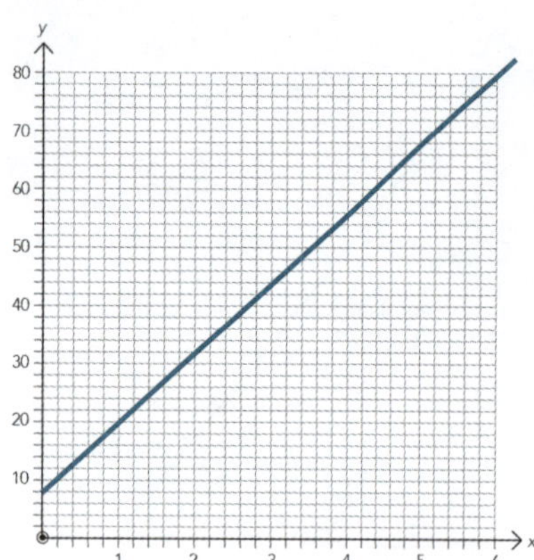

2 Calculating gradients (pp. 78–79)

1 m = 50

2 $m = \frac{1}{2}$

3 $m = -\frac{1}{2}$

4 m = -5

5 m = -4

6 $m = \frac{2}{5}$

7 $m = -\frac{1}{4}$

8 m = 8

Drawing straight lines — continuous data (pp. 80–89)

1 Plotting points using the equation (pp. 80–82)

1

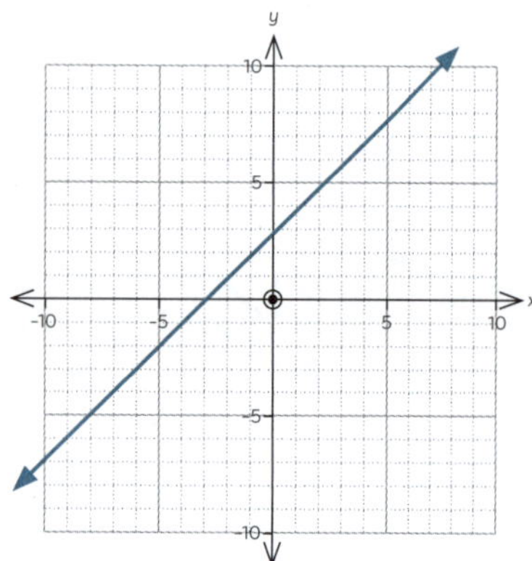

2

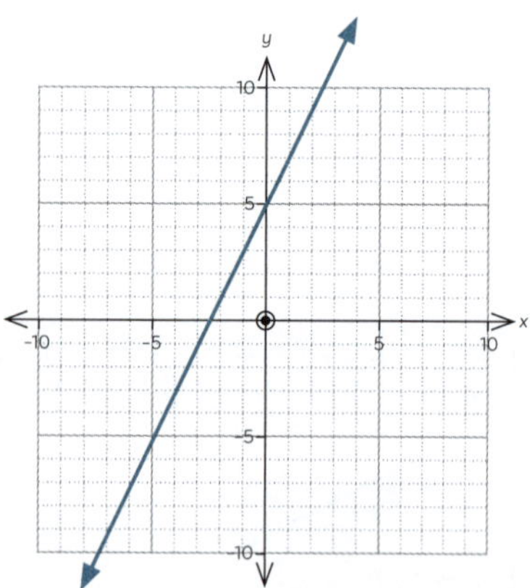

3

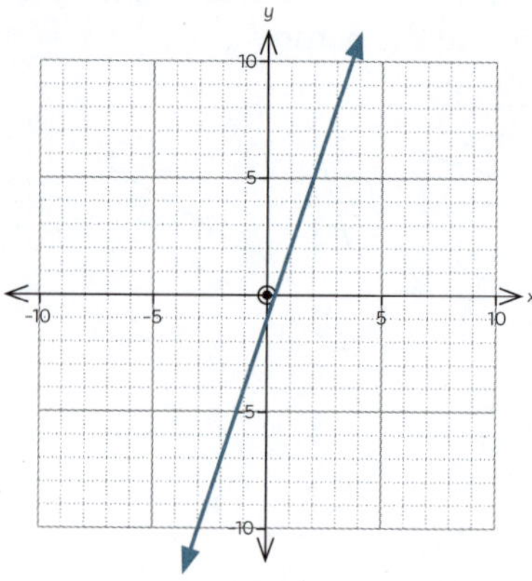

4

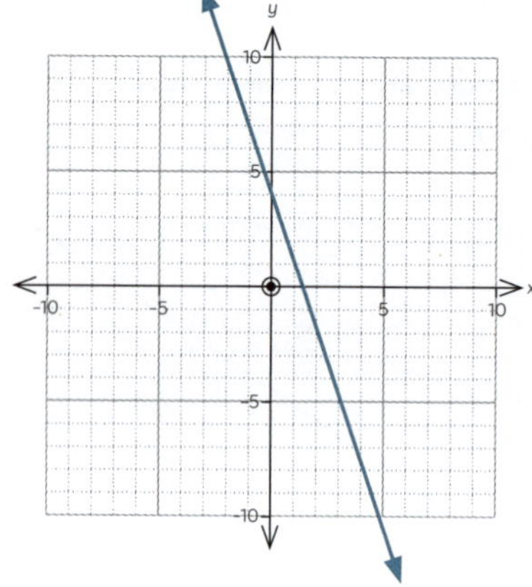

5

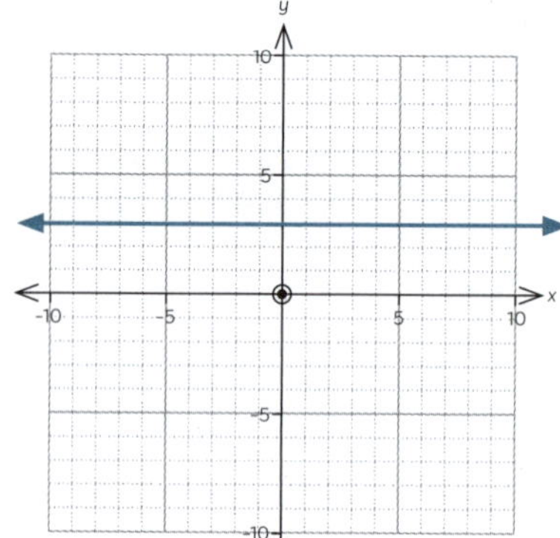

6

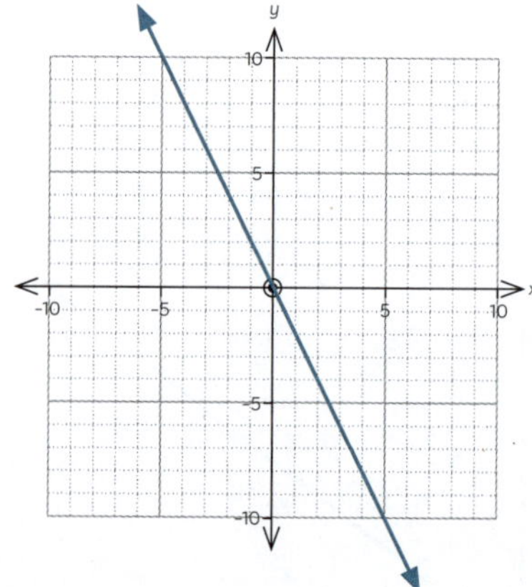

 ISBN: 9780170484084

2 Using the y-intercept and the gradient (pp. 83–85)

1 and 2

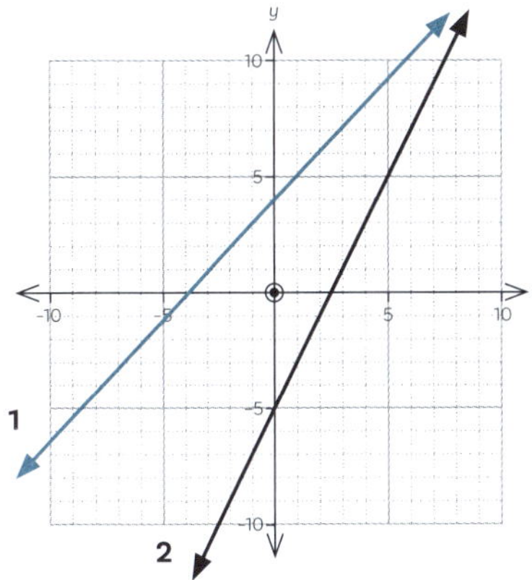

3 and 4

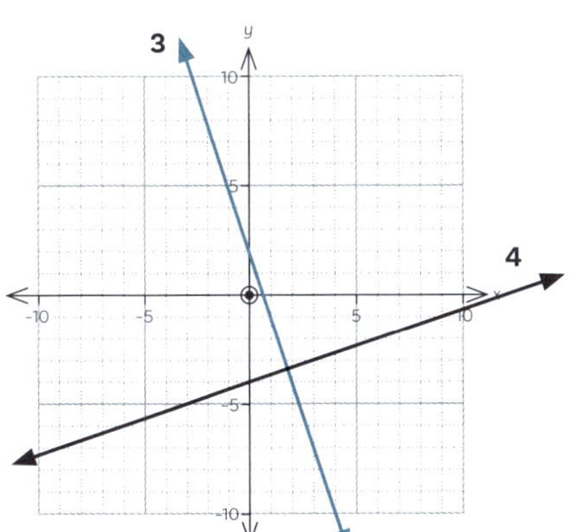

5 and 6

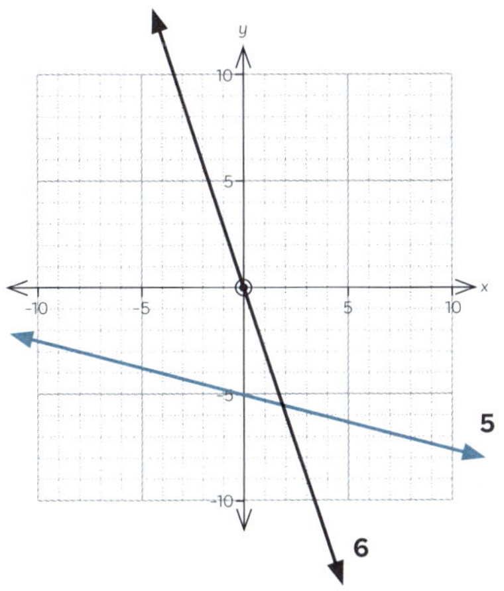

7 and 8

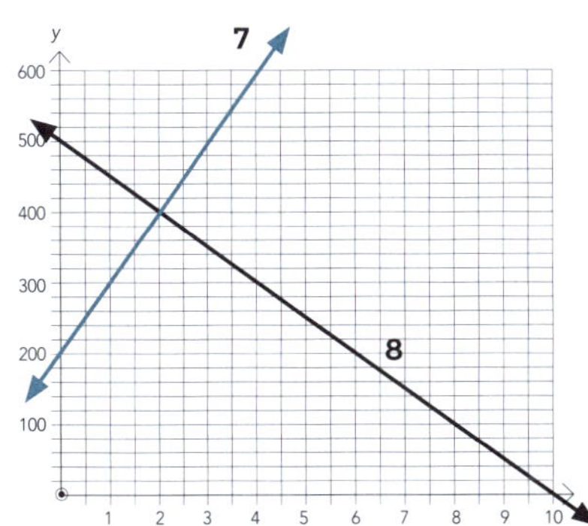

9 and 10

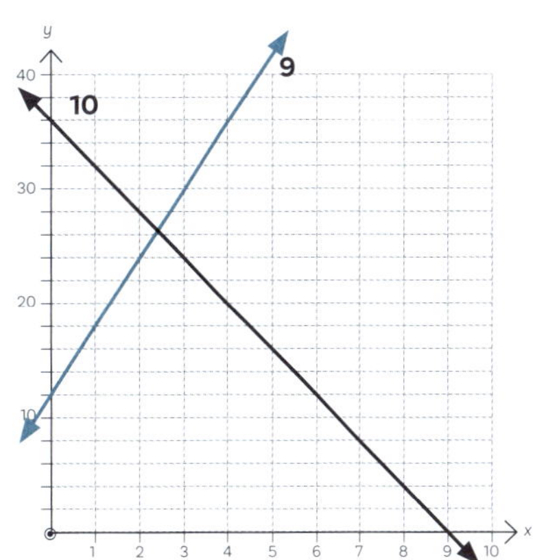

11 and 12

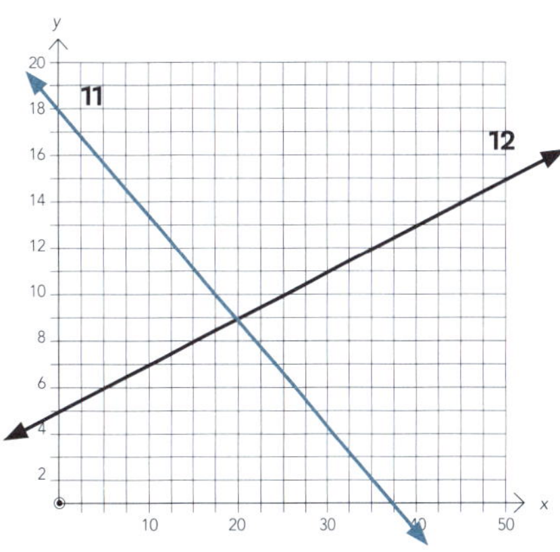

3 Using the x- and y-intercepts (pp. 86–87)

1 and 2

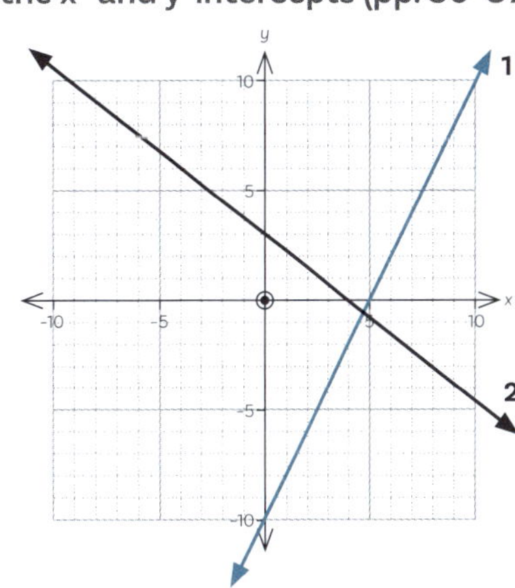

3 and 4

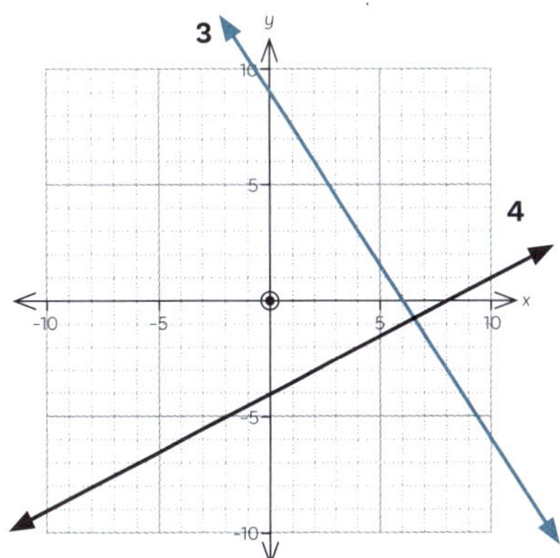

ISBN: 9780170484084

5 and 6

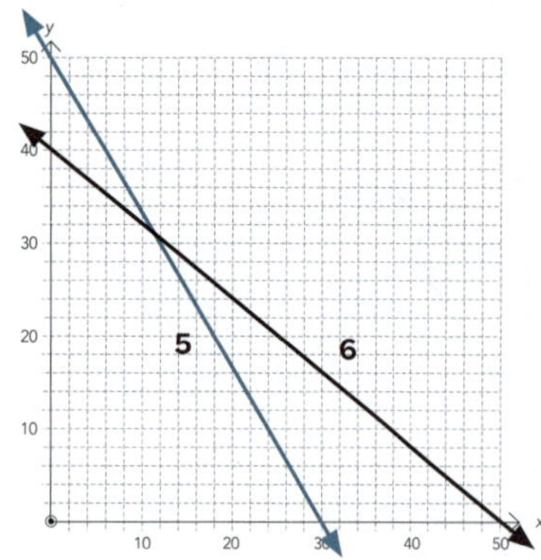

4 Horizontal and vertical lines (pp. 88–89)

1

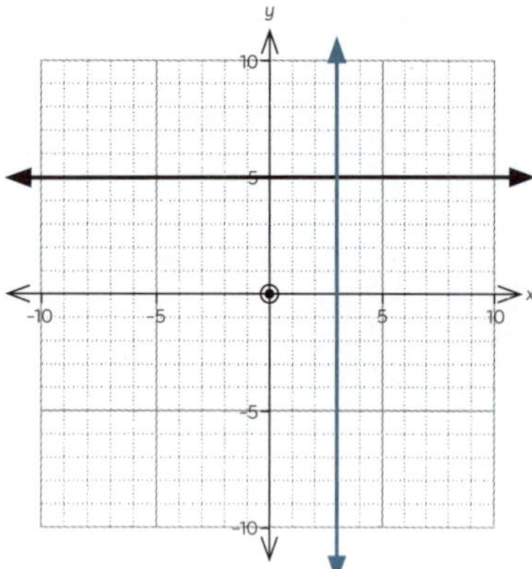

2

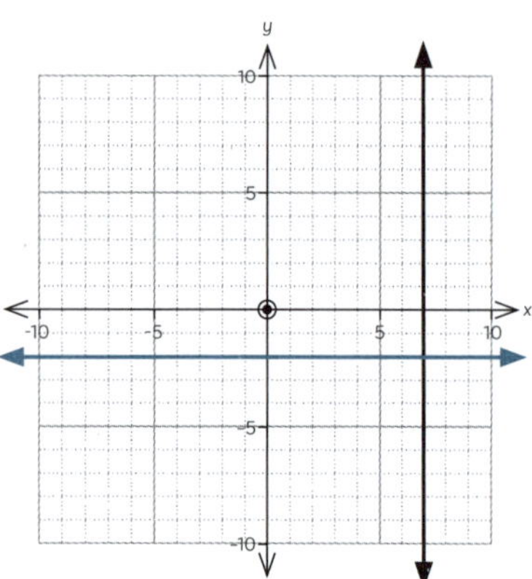

3

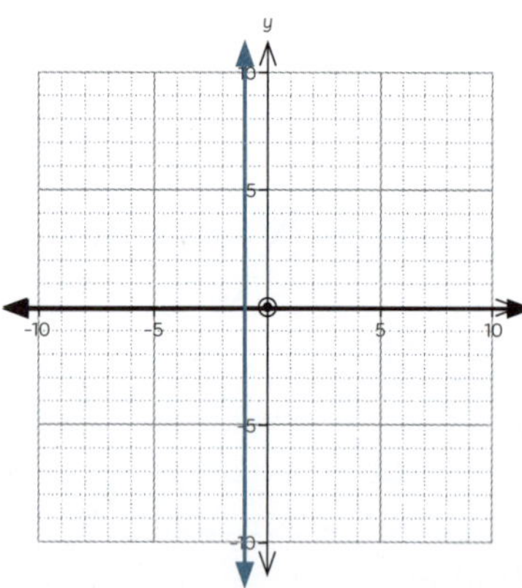

4

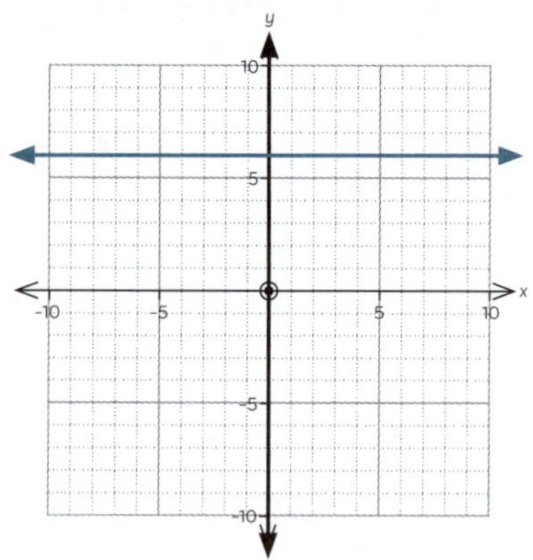

5

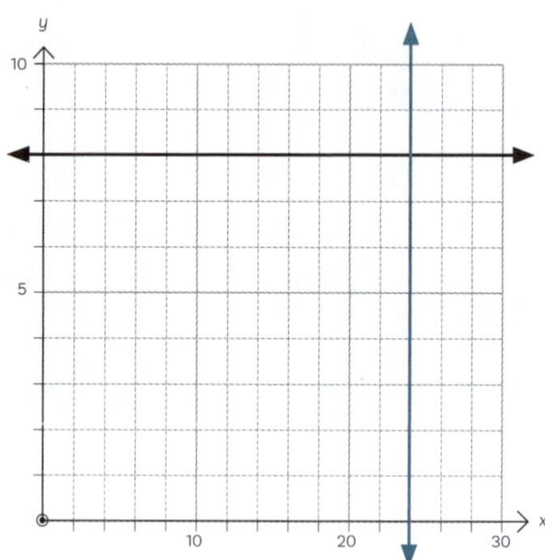

6

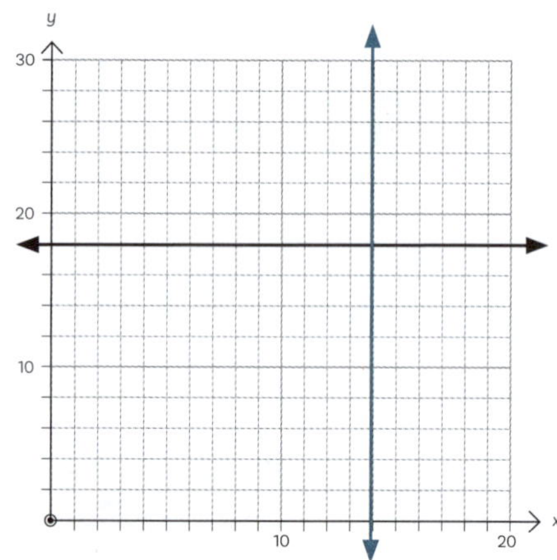

Writing equations from graphs (pp. 90–95)

1 Using the gradient and the y-intercept (pp. 90–93)

1 y-intercept = 3 Gradient = 2
$y = 2x + 3$

2 y-intercept = −3 Gradient = −1
$y = -x - 3$

3 y-intercept = 5 Gradient = −4
$y = -4x + 5$

4 y-intercept = −3 Gradient = $\frac{1}{2}$
$y = \frac{1}{2}x - 3$

5 $y = \frac{1}{2}x + 3$

6 $B = -5D + 80$

7 $y = -5x + 85$

8 $T = -\frac{1}{2}m + 38$

9 $A = \frac{1}{4}G + 12$

10 $y = -11x + 66$

11 $A = \frac{2}{5}G + 8$

12 $y = 10x + 16$

 ISBN: 9780170484084

2 Using two points (pp. 94–95)

1 $y = 2x + 3$
2 $y = \frac{1}{2}x + 7$
3 $y = x - 4$
4 $y = 2x + 5$
5 $y = -2x + 3$
6 $y = -x + 6$
7 $y = -\frac{1}{2}x + 1$
8 $y = -4x - 3$

GEOMETRY (pp. 96–130)

Naming angles (p. 96)

1 ∠CAB
2 ∠CBA
3 ∠a
4 ∠CBD
5 ∠BCD
6 ∠b

Types of angles (p. 97)

1 Obtuse angle
2 Reflex angle
3 Right angle
4 Acute angle
5 Straight angle
6 Reflex angle

Angle rules (pp. 98–100)

1 $y = 124°$ (∠s at a point = 360°)
2 $x = 129°$ (∠s on a line = 180°)
3 $a = 105°$ (vert opp ∠s are equal)
4 $h = 65°$ (∠s on a line = 180°)
5 $p = 158°$ (∠s at a point = 360°)
6 $g = 119°$ (vert opp ∠s are equal)
7 $f = 37°$ (vert opp ∠s are equal and ∠s at a point = 360° or ∠s on a line = 180°)
8 $e = 22°$ (∠s on a line = 180° or vert opp ∠s are equal)
9 $a = 55°$ (vert opp ∠s =)
$b = 85°$ (∠s on a line = 180°)
$c = 40°$ (∠s in Δ = 180°)
10 $a = 45°$ (vert opp ∠s =)
$b = 25°$ (∠s on a line = 180°)
$c = 110°$ (∠s in Δ = 180°)
$d = 120°$ (∠s at a point = 360°)
11 $a = 45°$ (∠s on a line = 180°)
$b = 60°$ (∠s at a point = 360°)
$c = 105°$ (ext ∠ of Δ = sum of int opp ∠s)
12 $a = 88°$ (vert opp ∠s =)
$b = 114°$ (ext ∠ of Δ = sum of int opp ∠s)
$c = 114° - 88°$
$= 26°$ (ext ∠ of Δ = sum of int opp ∠s)
13 $a = 40°$ (∠s on a line = 180°)
$b = 50°$ (∠s in Δ = 180°)
$c = 69°$ (vert opp ∠s =)
$d = 61°$ (∠s in Δ = 180°)

Parallel lines (pp. 101–102)

1 Alternate
2 Co-interior
3 Co-interior
4 Corresponding
5 Corresponding
6 Alternate
7 Co-interior
8 Corresponding

Calculating angles where parallel lines are involved (p. 102)

You may find other ways of doing these. Check with your teacher.

1 $a = 50°$ (co-int ∠s add to 180°, // lines)
$b = 65°$ (corr ∠s =, // lines)
$c = 65°$ (∠s on a line = 180°)
2 $a = 92°$ (co-int ∠s add to 180°, // lines)
$b = 36°$ (alt ∠s =, // lines)
$c = 128°$ (∠s in Δ = 180° and ∠s on a line = 180°)
3 $a = 109°$ (∠s in Δ = 180°)
$b = 109°$ (corr ∠s =, // lines)
$c = 26°$ (alt ∠s =, // lines)

Polygons (pp. 103–110)

Triangles (pp. 103–104)

1 $a = 50°$ (isos Δ, base ∠s =)
$b = 80°$ (∠s in Δ = 180°)
$c = 50°$ (alt ∠s =, // lines)
2 $a = 41°$ (co-int ∠s add to 180°, // lines)
$b = 56°$ (isos Δ, base ∠s =)
$c = 83°$ (∠s on a line = 180°)
3 $a = 66°$ (∠s on a line = 180°)
$b = 57°$ (isos Δ, base ∠s = and ∠s in Δ = 180°)
$c = 57°$ (isos Δ, base ∠s =)
$d = 123°$ (∠s on a line = 180°)
4 $a = 68°$ (alt ∠s =, // lines)
$b = 56°$ (isos Δ, base ∠s = and ∠s in Δ = 180°)
$c = 56°$ (isos Δ, base ∠s =)
5 ∠CBD = 60° (equilat Δ)
∴ $\theta = 81°$ (∠s on a line = 180°)
6 ∠CAB = 44° (isos Δ, base ∠s =)
∠DAC = 80°
∴ $\theta = 50°$ (isos Δ, base ∠s =)
7 ∠BCD = 64° (∠s on a line = 180°)
∠DBC = 52° (isos Δ, base ∠s =)
∴ $\theta = 33°$ (alt ∠s =, // lines)

Quadrilaterals (pp. 105–106)

1 $a = 71°$ (∠s of quad = 360°)
$b = 239°$ (∠s at a point = 360°)
$c = 55°$ (∠s of quad = 360°)
2 $a = 117°$ (alt ∠s =, // lines)
$b = 117°$ (opp ∠s of parallelogram =)
$c = 63°$ (∠s on a line = 180°)
$d = 63°$ (opp ∠s of parallelogram =)
3 $a = 114°$ (symmetry — opp ∠s of kite =)
$b = 27°$ (congruent Δs)
$c = 78°$ (∠s of quad = 360°)
4 $a = 124°$ (∠s on a line = 180°)
$b = 124°$ (base ∠s isos trapezium =)
$c = 56°$ (co-int ∠s add to 180°, // lines)
$d = 56°$(co-int ∠s add to 180°, // lines)
5 $a = 90°$ (diags of rhombus ⊥)
$b = 27°$ (isos Δ, base ∠s equal)
$c = 126°$ (co-int ∠s add to 180°, // lines)

Polygons in general (pp. 107–110)

Exterior and interior angles of polygons

1 112°
2 67°
3 141°
4 95°
5 110°
6 90°
7 265°
8 75°
9 273°
10 72°
11 36°
12 140°

ISBN: 9780170484084

The Theorem of Pythagoras (pp. 111–114)

Finding the length of the hypotenuse (pp. 111–112)

1 $x = 19.21$ m
2 $c = 9.434$ m
3 $x = 13$ cm
4 $x = 12.08$ cm
5 $x = 6.940$ m
6 $x = 42.43$ mm
7 $L = 91.53$ mm
8 $P = 104.8$ m

Finding the lengths of short sides (p. 113)

1 $x = 16.94$ mm
2 $x = 48.99$ mm
3 $x = 7.483$ cm
4 $L = 15.49$ km
5 $L = 45.92$ mm
6 $P = 0.5905$ km

Mixing it up (p. 114)

1 50.99 cm
2 122.1 mm
3 12 cm
4 6.788 cm
5 4.272 km
6 68.23 mm
7 61.81 cm
8 128.7 km
9 363.0 mm
10 0.5318 cm

Trigonometry (pp. 115–126)

Labelling sides (p. 115)

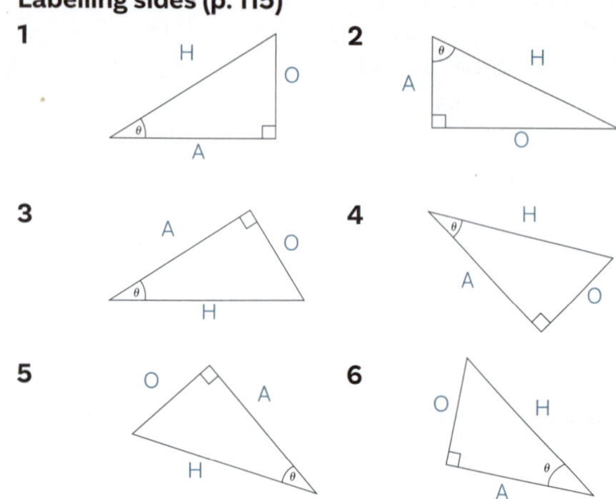

Finding short sides using sine and cosine (pp. 116–118)

1 11.87 cm
2 7.419 cm
3 32.48 mm
4 77.63 mm
5 3.011 m
6 8.940 km
7 398.1 km
8 321.9 mm
9 0.8954 m
10 40.38 cm

Finding long sides (hypotenuses) using sine and cosine (pp. 119–120)

1 26.42 cm
2 16.51 cm
3 138.2 mm
4 214.4 mm
5 6.055 m
6 30.42 km
7 1039 km

Finding sides using tangents (pp. 121–122)

1 8.748 cm
2 22.40 cm
3 120.9 mm
4 171.2 mm
5 4.052 m
6 28.93 km
7 439.3 km

Finding angles (pp. 123–124)

1 47.5°
2 29.7°
3 52.3°
4 49.1°
5 57.4°
6 65.0°
7 25.5°
8 54.6°
9 35.2°

Putting it together – with Pythagoras (pp. 125–126)

1 31.6°
2 9.793 cm
3 34.44 mm
4 49.1°
5 35.0 cm
6 6.66 cm
7 52.7°
8 4.61 cm
9 5.34 km
10 1.43 km
11 200 mm

3D shapes (pp. 127–128)

1 **a** HC = 53 cm **b** EC = 59.94 cm
c ∠ECH = 27.8° **d** ∠CHG = 31.9°
e ∠DHC = 58.1°

Bearings (pp. 129–130)

1 040°
2 108°
3 304°
4 248°
5 211°
6 337°
7 082°
8 158°
9

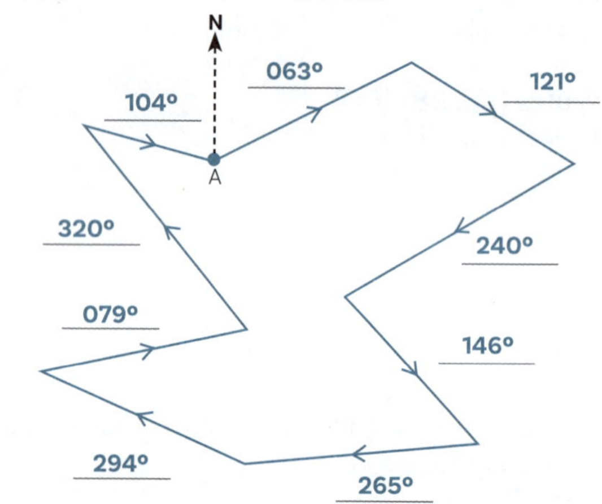

MEASUREMENT (pp. 131–147)

Unit conversion (pp. 131–132)

1 120 mm
2 3.56 kg
3 2.95 L
4 8900 m
5 3.094 t
6 250 mL
7 710 mg
8 0.003 km
9 280 kg
10 0.0186 km
11 0.095 L
12 0.298 kg
13 4.42 m
14 1.15 L
15 3.81 kg
16 6.42 km
17 1.322 t
18 5.817 m
19 13.465 km or 13 465 m
20 3.46 L or 3460 mL
21 5.5 t
22 28.43 km

Perimeter (pp. 133–135)

1 Figures with lines (p. 133)

1 40 mm
2 52 m
3 28 cm
4 12 cm
5 11.6 cm
6 55 mm

2 Circles – the circumference (p. 134)

1 25.13 mm
2 8.168 m
3 540.4 mm
4 50.27 cm
5 8.954 km
6 12.30 cm

ISBN: 9780170484084

3 Compound shapes (p. 135)

1	26 m	**2**	16.2 cm
3	38.85 cm	**4**	12 km
5	890.67 mm	**6**	644 mm

Area (pp. 136–141)

1 Triangles and quadrilaterals (pp. 136–137)

Shape	Area	Shape	Area
Rectangle (h, b)	area = base x height	Parallelogram and rhombus (h, a, b)	area = base x vertical height
Triangle (a, h, c, b)	area = $\frac{1}{2}$ x base x height	Trapezium (d, c, h, a, b)	area = $\frac{(a+b)}{2}$ x height

1	117 m^2	**2**	0.8084 cm^2
3	57 cm^2	**4**	240 km^2
5	11.5 km^2	**6**	64.26 cm^2
7	25.92 cm^2	**8**	56 cm^2
9	3136 cm^2	**10**	27.5 cm^2
11	6048 cm^2	**12**	9 m

2 Circles (pp. 138–139)

1	4.988 cm^2	**2**	31.17 km^2
3	0.9852 m^2	**4**	24 660 cm^2
5	36.95 m^2	**6**	7389.03 cm^2
7	66.48 km^2	**8**	6.000 cm
9	38.00 m	**10**	4.233 cm^2
11	0.2300 m	**12**	1.744 m^2
13	61.26 cm^2	**14**	x = 10.00 cm

3 Compound shapes (p. 140)

1	369 cm^2	**2**	12.87 m^2
3	50 920 mm^2	**4**	4484 cm^2
5	33.76 m^2	**6**	0.6475 km^2

4 Shapes with holes (p.141)

1	8.44 m^2	**2**	5817.5 mm^2
3	88.3 m^2	**4**	144.66 cm^2
5	202.71 km^2	**6**	1.57 m^2

Volume (pp. 142–144)

1 Cuboids (p. 142)

1	13 500 mm^3	**2**	55.56 m^3
3	451 500 mm^3	**4**	45.9 cm^3

2 Cylinders (p. 143)

1	1460 cm^3	**2**	650.2 cm^3
3	1109 cm^3	**4**	74.12 m^3
5	15 540 cm^3	**6**	1.059 m^3

3 Triangular prisms (p. 144)

1	959 100 mm^3	**2**	101.6 cm^3
3	1428 m^3	**4**	21 240 cm^3

Surface area (pp. 145–146)

1 Cuboids (p. 145)

1	37.50 cm^2	**2**	97.02 cm^2
3	35 440 mm^2	**4**	95.1 cm^2

2 Cylinders, prisms and pyramids (p. 146)

1	240 m^2	**2**	471.2 cm^2
3	360 cm^2	**4**	352 cm^2
5	298.0 mm^2	**6**	144 cm^2
7	61.32 cm^2		

STATISTICS AND PROBABILITY (pp. 148–164)

Probability (pp. 148–159)

The range of values for probabilities (p. 148)

0	impossible, no chance, no way
0.1	very unlikely, slight chance, possible
0.2	improbable, possible, unlikely
0.3	improbable, maybe, possible, unlikely
0.4	maybe, possible, unlikely
0.5	maybe, possible
0.6	likely, probable
0.7	likely, probable
0.8	very likely, probable
0.9	very likely, extremely likely
1.0	definite, a sure thing

Using numbers for writing probability (p. 149)

1 $\frac{7}{16} = 0.4375$ $\frac{1}{3} = 0.\dot{3}$ More likely: $\frac{7}{16}$

2 $\frac{3}{4} = 0.75$ $\frac{8}{11} = 0.\dot{7}\dot{2}$ More likely: $\frac{3}{4}$

3 $\frac{17}{18} = 0.9\dot{4}$ $\frac{19}{20} = 0.95$ More likely: $\frac{19}{20}$

4 $\frac{13}{24} = 0.541\dot{6}$ $\frac{27}{50} = 0.54$ More likely: $\frac{13}{24}$

5 $\frac{11}{20} = 0.55$ $\frac{7}{13} = 0.5385$ More likely: $\frac{11}{20}$

6 $\frac{6}{17} = 0.3529$ $\frac{3}{8} = 0.375$ More likely: $\frac{3}{8}$

7 ✓ 0.009 means there is a very small chance that it will be sunny.

8 ✗ Probabilities cannot be negative, must be between 0 and 1.

9 ✗ Probabilities cannot be greater than 1.

10 ✓ $\frac{1}{200}$ is a valid way of writing a probability, but it is very unlikely.

ISBN: 9780170484084

Ways of calculating probabilities (p. 150)

1	$0.1\dot{6}$ or $\frac{1}{6}$	**2**	0.5 or $\frac{1}{2}$
3	0.5 or $\frac{1}{2}$	**4**	$0.8\dot{3}$ or $\frac{5}{6}$
5	0	**6**	1
7	0.25 or $\frac{1}{4}$	**8**	0.5 or $\frac{1}{2}$
9	0.0769 or $\frac{4}{52}$	**10**	0.0385 or $\frac{2}{52}$
11	0.1538 or $\frac{8}{52}$	**12**	0.1154 or $\frac{6}{52}$
13	0.9231 or $\frac{48}{52}$	**14**	0.8462 or $\frac{44}{52}$
15	$0.291\dot{6}$ or $\frac{7}{24}$	**16**	$0.708\dot{3}$ or $\frac{17}{24}$
17	$0.41\dot{6}$ or $\frac{10}{24}$	**18**	$0.58\dot{3}$
19	$0.791\dot{6}$ or $\frac{19}{24}$	**20**	$0.791\dot{6}$ or $\frac{19}{24}$
21	0	**22**	1

Expected number of outcomes (p. 151)

1	40 heads	**2**	30 times
3	79 or 80 people	**4**	30 or 31 boys
5	1 or 2 eggs	**6**	1 hand

Combining probabilities – probability trees (pp. 152–156)

1 a

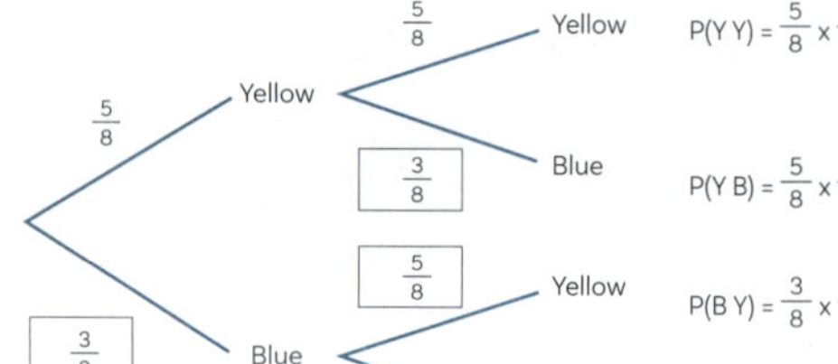
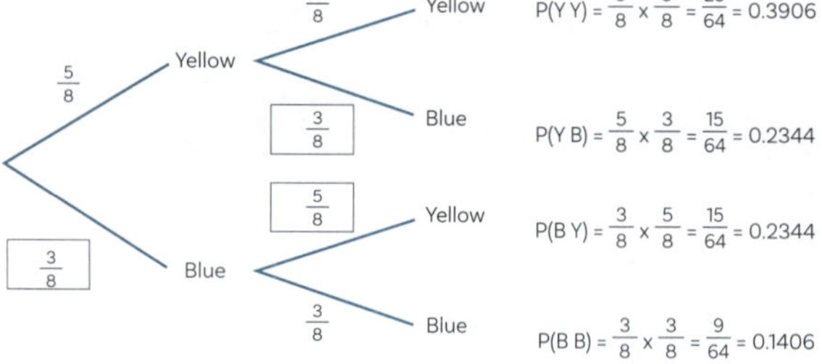

b $\frac{25}{64}$ or 0.3906 **c** $\frac{30}{64}$ or 0.4688

d $\frac{34}{64}$ or 0.5313 **e** $\frac{9}{64}$ or 0.1406

f $\frac{39}{64}$ or 0.6094 **g** 78 or 79

h 121 or 122

2 a

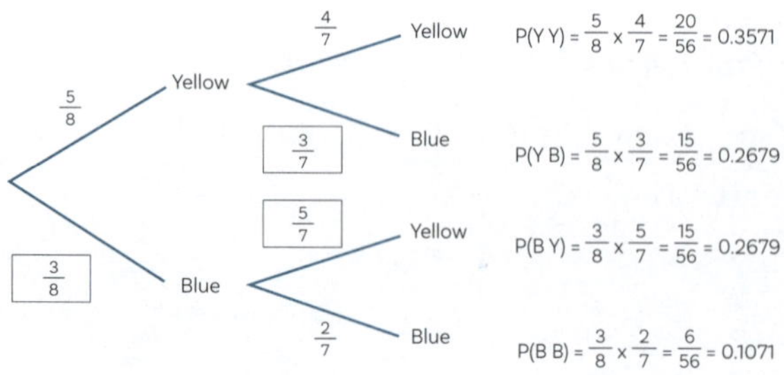

b $\frac{20}{56}$ or 0.3571 **c** $\frac{30}{56}$ or 0.5357

d $\frac{26}{56}$ or 0.4643 **e** $\frac{6}{56}$ or 0.1071

f $\frac{36}{56}$ or 0.6429 **g** $\frac{30}{56}$ or 0.5357

h 1

3 a

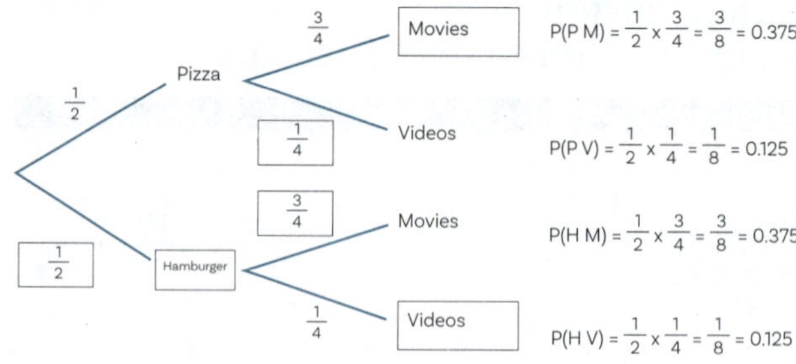

b $\frac{6}{8}$ or 0.75 **c** $\frac{1}{8}$ or 0.125

d $\frac{3}{8}$ or 0.375 **e** $\frac{7}{8}$ or 0.875

f $\frac{5}{8}$ or 0.625 **g** $\frac{3}{8}$ or 0.375

h $\frac{1}{8}$ or 0.125

Two-way tables (pp. 157–159)

1

	Year 9	Year 10	Totals
Went to camp	51	67	**118**
Didn't go to camp	23	9	**32**
Totals	**74**	**76**	**150**

a $\frac{118}{150}$ or 0.79 **b** $\frac{67}{150}$ or 0.45

c $\frac{51}{74}$ or 0.69 **d** $\frac{118}{150}$ x $\frac{117}{149}$ = 0.62

2

	Plays a musical instrument	Doesn't play a musical instrument	Totals
Left handed	6	**5**	**11**
Right handed	**54**	29	**83**
Totals	**60**	**34**	**94**

a $\frac{11}{94}$ x 100 = 11.7% **b** $\frac{54}{94}$ or 0.57

c $\frac{29}{83}$ or 0.35 **d** $\frac{60}{94}$ x $\frac{59}{93}$ = 0.40

3

	Had been skiing	Had not been skiing	Totals
Boys	11	**28**	**39**
Girls	16	**29**	**45**
Totals	**27**	**57**	**84**

a $\frac{27}{84}$ or 0.3214 **b** $\frac{11}{84}$ or 0.1310

c $\frac{16}{45}$ or $0.3\dot{5}$ **d** $\frac{27}{84}$ x $\frac{26}{83}$ = 0.1007

ISBN: 9780170484084

4

	Primary	Secondary	Totals
Dinner at a table	0.27	0.30	0.57
Dinner not at a table	0.18	0.25	0.43
Totals	0.45	0.55	1

a $0.57 \times 1200 = 684$ **b** $0.18 \times 1200 = 216$

c $\frac{216}{540} = 0.4$ **d** 984

Data analysis (pp. 160–164)

Measures of centre (pp. 160–162)

1 Mean = 2.6
Median = 2.5
Mode = 1

2 Mean = $17.\dot{7}\dot{2}$ (2 dp)
Median = 19
Mode = 20

3 Mean = 30.65
Median = 28.65
Mode = no mode

4 Mean = $13.\dot{5}\dot{4}$ (2 dp)
Median = 7
Mode = 2
The mean and median are different. There is one large value (89) which affects the mean but not the median.

5 Mean = 4.225 (3 dp)
Median = 4.15
Mode = no mode
The mean and median are similar. The data is symmetrically distributed with no unusual values.

6 **a** 18 **b** 15

7

Set A	2	18	6
Set B	13	12	1
Set C	16	4	6

Measures of spread (pp. 163–164)

1 Minimum = 5
LQ = 11
Median = 17
UQ = 21
Maximum = 25
Range = 20
Interquartile range = 10

2 Minimum = 0
LQ = 1
Median = 2.5
UQ = 5
Maximum = 8
Range = 8
Interquartile range = 4

3 Minimum = 7
LQ = 10
Median = 12.5
UQ = 15.5
Maximum = 21
Range = 14
Interquartile range = 5.5

4

Mean		Median	✓	IQR	✓	Range	

5

Mean		Median	✓	IQR	✓	Range	

6

Mean		Median	✓	IQR	✓	Range	

7 When the largest or smallest number in a data set is changed, the **median** and the **interquartile range** stay the same, but the **mean** and the **range** become different.

Practice sets (pp. 165–174)

Set 1 (pp. 165–166)

1 $7x^2yz - 2x^2y - 2xy$

2 $-\frac{5}{8}$ or -0.625

3 27 mL

4 $23x - 4 - 15x^2$ or $-15x^2 + 23x - 4$

5 $\tan x = \frac{40}{65}$
$x = 31.6°$

6 $A = 6 \times \frac{9 + 15}{2}$
$= 72\text{ cm}^2$

7 $\frac{1}{2}$

8 Internal angles of a pentagon add to 540°.
$x = 540° - 98° - 90° - 30° - 240°$
$= 82°$

9 $3x(4xy - x^4 + 3)$

10 **a**

	Year 9	Year 10	Total
Drama	24	27	51
Kapa haka	36	48	84
Total	60	75	135

b 0.4 or $\frac{24}{60}$

c $0.\dot{6}2$ or $\frac{84}{135}$

Set 2 (pp. 167–168)

1 $(x - 10)(x - 1)$

2 $\frac{11}{15} = 0.7\dot{3}$

$\therefore$ an event with a probability of $\frac{11}{15}$ is more likely.

3 $AC = \sqrt{15^2 + 15^2}$

$= 21.21$ cm (2 dp)

4 $\frac{-4y^2z}{x^3}$

5 101.92 kg

6 $y = 2x + 3$

7 $V = \pi r^2h$

$= \pi \times 37^2 \times 56$

$= 240\ 800$ mm^3 (4 sf)

8 $\frac{4x^2}{3y} \div \frac{x}{y} = \frac{4x^2}{3y} \times \frac{y^2}{x}$

$= \frac{4xy}{3}$

9 51.4°

10 $4(x - 5) = 23 - 6x$

$4x - 20 = 23 - 6x$

$10x = 43$

$x = 4.3$

11 Area $= 58 \times 165$

$= 9\ 570$ cm^2

$= 0.957$ m^2

12 $\frac{5}{x-1} \times \frac{3x+2}{3x+2} + \frac{2}{3x+2} \times \frac{x-1}{x-1}$

$= \frac{5(3x+2) + 2(x-1)}{(x-1)(3x+2)}$

$= \frac{17x + 8}{(x-1)(3x+2)}$

Set 3 (pp. 169–170)

1 $y = 4x + 3$

2 $\frac{3x+5}{4} + x = \frac{3x+5}{4} + \frac{4x}{4}$

$= \frac{7x+5}{4}$

3 184 mm

4 $10x + 15 \leq 24 - 6x$

$16x \leq 9$

$x \leq \frac{9}{16}$

5 $H = \sqrt{76.32^2 - 65^2}$

$= 40$ cm (0 dp)

6 $-2(-3x3y)^2 = -2(9x^6y^2)$

$= -18x^6y^2$

7 $1\frac{1}{8}$ cups of cheese

8 $\cos x = \frac{25}{65}$

$x = 67.4°$ (1 dp)

9 24%

10 a

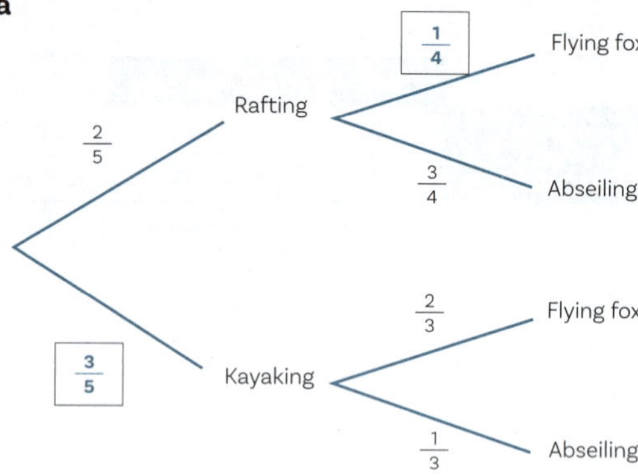

b $P = \frac{2}{5} \times \frac{3}{4}$

$= 0.3$

c $P = \frac{2}{5} \times \frac{3}{4} + \frac{3}{5} \times \frac{1}{3}$

$= \frac{6}{20} \times \frac{3}{15}$

$= 0.5$

Set 4 (pp. 171–172)

1

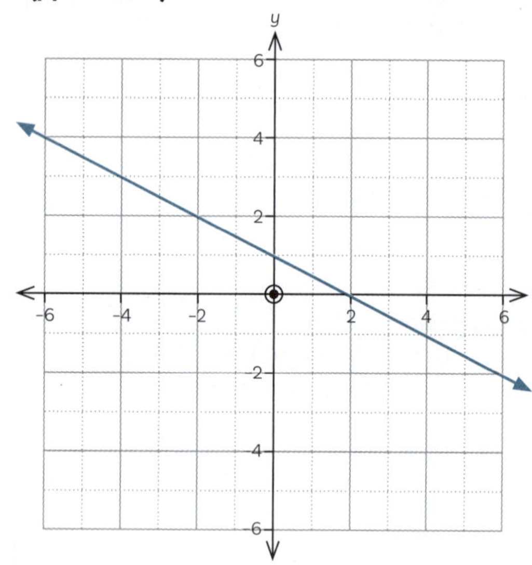

2 $4x^8y^{18}$

3 $A = \frac{1}{2}\pi \times 15^2$

$= 353.43$ cm^2 (2 dp)

4 $(x + 2)(x - 7)$

5 $\sin 42° = \frac{45}{AC}$

$AC = \frac{45}{42 \sin°}$

$AC = 67.25$ mm (2 dp)

6 $V = 3^3 - \frac{1}{3} \times \pi \times 3^3$

$= 27 - 9\pi$

7 72 cm

8 $\angle DFE = 52°$

(ext $\angle\ \Delta$ = sum int opp $\angle$s)

$\angle x = 128°$

(co-int $\angle$s = 180°, // lines)

9 $\frac{x}{2} + 5 - x = 7$ (x 2)

$x + 10 - 2x = 14$

$x = -4$

 ISBN: 9780170484084

10 $24 = 2 \times 2 \times 2 \times 3$

11 315°

12 $77.52

Set 5 (pp. 173–174)

1 $2A = ah + bh$

$ah = 2A - bh$

$a = \frac{2A - bh}{h}$ or $\frac{2A}{h} - b$

2 Any length between 6 and 6.3 cm.

3 54

4 $\tan 35° = \frac{37}{BC}$

$BC = \frac{37}{\tan 35°}$

BC = 52.84 mm (2 dp)

5 $6x - 15 - 4x + x^2 = x^2 + 2x - 15$

6 $x^3 = -125$

$x = -5$

7 7110

8 $HC = \sqrt{64^2 + 14^2}$

$= 65.513$

$HB = \sqrt{25^2 + 65.513^2}$

= 70.12 cm (2 dp)

9 $\text{Price} = \frac{\$112.70}{1.15}$

= $98

10 $7 - 2x = 7x + 74$

$-9x = 67$

$x = -7.\dot{4}$

11 $y = -2x + 5$

12 6490 cm^3

 ISBN: 9780170484084